4th Social Media Mining for Health Applications Workshop & Shared Task 2019

Florence, Italy
2 August 2019

ISBN: 978-1-5108-9113-5

Printed by Curran Associates, Inc. (2019)

For permission requests, please contact the Association for Computational Linguistics
at the address below.

Association for Computational Linguistics
209 N. Eighth Street
Stroudsburg, Pennsylvania 18360

Phone: 1-570-476-8006
Fax: 1-570-476-0860

acl@aclweb.org

Additional copies of this publication are available from:

Curran Associates, Inc.
57 Morehouse Lane
Red Hook, NY 12571 USA
Phone: 845-758-0400
Fax: 845-758-2633
Email: curran@proceedings.com
Web: www.proceedings.com

ACL 2019

**Social Media Mining for Health Applications (#SMM4H)
Workshop & Shared Task**

Proceedings of the Fourth Workshop

August 2, 2019
Florence, Italy

Preface

Welcome to the 4th Social Media Mining for Health Applications Workshop and Shared Task - #SMM4H 2019.

The total number of users of social media continues to grow worldwide, resulting in the generation of vast amounts of data. Popular social networking sites such as Facebook, Twitter and Instagram dominate this sphere. According to estimates, 500 million tweets and 4.3 billion Facebook messages are posted every day [1]. The latest Pew Research Report [2], nearly half of adults worldwide and two-thirds of all American adults (65%) use social networking. The report states that of the total users, 26% have discussed health information, and, of those, 30% changed behavior based on this information and 42% discussed current medical conditions. Advances in automated data processing, machine learning and NLP present the possibility of utilizing this massive data source for biomedical and public health applications, if researchers address the methodological challenges unique to this media.

In its fourth iteration, the #SMM4H workshop takes place in Florence, Italy, on August 2, 2019, and is co-located with the annual meeting of the Association of Computational Linguistic (ACL). Following on the success of our Workshops and accompanying shared tasks on the topic that were hosted at the Pacific Symposium in Biocomputing (PSB) in 2016, at the AMIA Annual Conference in 2017, and at the EMNLP conference in 2018, this workshop aims to provide a forum for the ACL community members to present and discuss NLP advances specific to social media use in the particularly challenging area of health applications, with a special focus given to automatic methods for the collection, extraction, representation, analysis, and validation of social media data for health informatics.

We received very high quality submissions for the workshop and selected only 8 articles for long presentations (Workshop Acceptance Rate: 54%) and 17 for short talks or posters presentations. As for the previous years, we ran in parallel to the workshop a shared task with a particular interest on social media mining for pharmacovigilance. For this fourth execution of the #SMM4H shared tasks, we challenged the community with two different problems involving annotated user posts from Twitter (tweets). The first problem focuses on performing pharmacovigilance from social media data during a series of three subtasks inviting the participants to extract and normalize tweets mentioning adverse effects of drugs. The second problem explores the generalizability of predictive models through a task of automatic classification of tweets with personal health experience mentions in multiple contexts. With a total of 34 teams registered and 19 teams having submitted a run, we confirm a growing interest of the community for health mining in social media data.

This year, we standardized the submission process during the shared task using the web platform Codalab[3]. We believed this helped improved the reproducibility of the experiments. Acting as a central hub, the data are easily distributed to the research community at large and, as the Codalab website remains active even after the competition, new teams can upload their submissions and be automatically evaluated to compare their results with the official results of the challenge.

Next year, we will go further to improve reproducibility by allowing participants to upload their code and models directly in Codalab for evaluation, a change which will guarantee a fair competition and the dissemination of the technical characteristics of the systems. Another important change this year that will impact future iterations of the workshop is an open call to the community for shared task proposals, ensuring that our workshop continues to address the main problems and challenges in this growing field

---

[1] Team Gwava. "How Much Data is Created on the Internet each Day?" 2016, Available online at `https://www.gwava.com/blog/internet-data-created-daily`. [Accessed: 03-Jan-2017].

[2] Pew Research Center. "Social Media Fact Sheet". 2017. Available online at `http://www.pewinternet.org/fact-sheet/social-media/`. [Accessed: 03-Mar-2017].

[3] CodaLab is free and open-source, available at `https://competitions.codalab.org/`.

of health mining for social media.

The organizing committee would like to thank the program committee, consisting of 13 researchers, for their thoughtful input on the submissions, as well as the organizers of the ACL for their support and management. Finally, a huge thanks to all authors who submitted a paper to the workshop or participated in the shared tasks; this workshop would not have been possible without them and their hard work.

Graciela, Davy, Abeed, Arjun, Ashlynn, Michael

Table of Contents

Conference Program

Friday, August 2, 2019

9:00–9:10   *Introduction*
Graciela Gonzalez-Hernandez

9:10–9:30   *Extracting Kinship from Obituary to Enhance Electronic Health Records for Genetic Research*
Kai He, Jialun Wu, Xiaoyong Ma, Chong Zhang, Ming Huang, Chen Li and Lixia Yao

9:30–9:50   *Lexical Normalization of User-Generated Medical Text*
Anne Dirkson, Suzan Verberne and Wessel Kraaij

9:50–10:10   *Overview of the Fourth Social Media Mining for Health (SMM4H) Shared Tasks at ACL 2019*
Davy Weissenbacher, Abeed Sarker, Arjun Magge, Ashlynn Daughton, Karen O'Connor, Michael J. Paul and Graciela Gonzalez-Hernandez

10:30–11:30   ***Coffee Break and Poster Session***

11:30–11:50   *MedNorm: A Corpus and Embeddings for Cross-terminology Medical Concept Normalisation*
Maksim Belousov, William G. Dixon and Goran Nenadic

11:50–12:00   *Passive Diagnosis Incorporating the PHQ-4 for Depression and Anxiety*
Fionn Delahunty, Robert Johansson and Mihael Arcan

12:00–14:00   ***Lunch***

14:00–14:20   *HITSZ-ICRC: A Report for SMM4H Shared Task 2019-Automatic Classification and Extraction of Adverse Effect Mentions in Tweets*
Shuai Chen, Yuanhang Huang, Xiaowei Huang, Haoming Qin, Jun Yan and Buzhou Tang

14:20–14:40   *KFU NLP Team at SMM4H 2019 Tasks: Want to Extract Adverse Drugs Reactions from Tweets? BERT to The Rescue*
Zulfat Miftahutdinov, Ilseyar Alimova and Elena Tutubalina

14:40–15:00   *Approaching SMM4H with Merged Models and Multi-task Learning*
Tilia Ellendorff, Lenz Furrer, Nicola Colic, Noëmi Aepli and Fabio Rinaldi

15:00–16:00   ***Coffee Break and Poster Session***

16:00–16:20   *Identifying Adverse Drug Events Mentions in Tweets Using Attentive, Collocated, and Aggregated Medical Representation*
Xinyan Zhao, Deahan Yu and V.G.Vinod Vydiswaran

16:20–16:40   *Correlating Twitter Language with Community-Level Health Outcomes*
Arno Schneuwly, Ralf Grubenmann, Séverine Rion Logean, Mark Cieliebak and Martin Jaggi

16:40–16:50   *Affective Behaviour Analysis of On-line User Interactions: Are On-line Support Groups More Therapeutic than Twitter?*
Giuliano Tortoreto, Evgeny Stepanov, Alessandra Cervone, Mateusz Dubiel and Giuseppe Riccardi

17:00–17:30   **Discussion and Conclusion**

Poster Sessions

Transfer Learning for Health-related Twitter Data
Anne Dirkson and Suzan Verberne

NLP@UNED at SMM4H 2019: Neural Networks Applied to Automatic Classifications of Adverse Effects Mentions in Tweets
Javier Cortes-Tejada, Juan Martinez-Romo and Lourdes Araujo

Detecting and Extracting of Adverse Drug Reaction Mentioning Tweets with Multi-Head Self Attention
Suyu Ge, Tao Qi, Chuhan Wu and Yongfeng Huang

Deep Learning for Identification of Adverse Effect Mentions In Twitter Data
Paul Barry and Ozlem Uzuner

Using Machine Learning and Deep Learning Methods to Find Mentions of Adverse Drug Reactions in Social Media
Pilar López Úbeda, Manuel Carlos Díaz Galiano, Maite Martin and L. Alfonso Urena Lopez

Towards Text Processing Pipelines to Identify Adverse Drug Events-related Tweets: University of Michigan @ SMM4H 2019 Task 1
V.G.Vinod Vydiswaran, Grace Ganzel, Bryan Romas, Deahan Yu, Amy Austin, Neha Bhomia, Socheatha Chan, Stephanie Hall, Van Le, Aaron Miller, Olawunmi Oduyebo, Aulia Song, Radhika Sondhi, Danny Teng, Hao Tseng, Kim Vuong and Stephanie Zimmerman

Neural Network to Identify Personal Health Experience Mention in Tweets Using BioBERT Embeddings
Shubham Gondane

Extracting Kinship from Obituary to Enhance
Electronic Health Records for Genetic Research

Kai He[1,2], Jialun Wu[1,2], Xiaoyong Ma[1,2], Chong Zhang[1,2],
Ming Huang[3], Chen Li[1,2*], Lixia Yao[3,*]

[1]School of Electronic and Information Engineering,
Xi'an Jiaotong University, Xi'an, China, 710049
[2]Shanxi Province Key Laboratory of Satellite and Terrestrial Network
Technology Research and Development, Xi'an Jiaotong University, Xi'an, China
[3]Department of Health Sciences Research, Mayo Clinic,
Rochester MN, USA, 55905
{hk52025804, andylun96, pixie1997, zc6063}@stu.xjtu.edu.cn
{Huang.ming, Yao.Lixia}@mayo.edu
cli@xjtu.edu.cn

Abstract

Claims database and electronic health records database do not usually capture kinship or family relationship information, which is imperative for genetic research. We identify online obituaries as a new data source and propose a special named entity recognition and relation extraction solution to extract names and kinships from online obituaries. Built on 1,809 annotated obituaries and a novel tagging scheme, our joint neural model achieved macro-averaged precision, recall and F measure of 72.69%, 78.54% and 74.93%, and micro-averaged precision, recall and F measure of 95.74%, 98.25% and 96.98% using 57 kinships with 10 or more examples in a 10-fold cross-validation experiment. The model performance improved dramatically when trained with 34 kinships with 50 or more examples. Leveraging additional information such as age, death date, birth date and residence mentioned by obituaries, we foresee a promising future of supplementing EHR databases with comprehensive and accurate kinship information for genetic research.

1 Introduction

Kinship or family relationship is important for genetic research, particularly for understanding trait and disease heritability, predicting individual disease susceptibility, and developing personalized medicine (Chatterjee et al., 2016). Human genetics started by analyzing pedigrees and twins to understand the roles of heredity and environment in the manifestation of physiological traits and diseases. With the rising of genomics, Electronic Health Records (EHRs) and their integration through biobank, kinship information, if available, can largely augment latest high-throughput computational technologies such as deep phenotyping from medical records (Robinson, 2012) and phenome-wide association study (PheWAS, Denny et al., 2010), and accelerate population-based genetic research (Mayer et al., 2014; Polderman et al., 2015). Unfortunately, neither EHR systems nor claims databases capture kinship information systematically.

A few studies have investigated disease heritability based on inferred kinship information. For example, Wang et al. selected 128,989 families of 481,657 individuals from a large claims database covering 1/3 of the US population, by selecting policyholders and their dependents (e.g., spouse and children) who were on file for at least 6 years, to estimate 149 diseases' heritability and familial environmental patterns (Wang et al., 2017). Similarly, Polubriaginof and colleagues performed a multi-center study based on 3,550,598 patients' medical records from three EHR systems in New York City and used emergency contact information to build more than 595,000 pedigrees, in order to compute the heritability of 500 disease phenotypes (Polubriaginof et al., 2018).

Proceedings of the 4th Social Media Mining for Health Applications (#SMM4H) Workshop & Shared Task, pages 1–10
Florence, Italy, August 2, 2019. ©2019 Association for Computational Linguistics

However, these studies relied on indirect sources to infer kinship information, which are incomplete and error-prone. First, both the dependents defined by medical insurance and the emergency contacts submitted to EHR systems by patients do not guarantee biological relationships. They do not distinguish adopted relationships or step relationships created through re-marriage from biological relationships. Second, dependents or emergency contact only represents a small portion of a person's whole family relationships. The 2010 Affordable Care Act allows young adults up to 26 to remain on their parents' health insurance plans. Before that, dependent children often "aged out" of their parents' health plan at age 19, or 22 if they were full-time students. Thus adult children older than those ages cannot be identified from claims data. In addition, if married couple work and receive medical insurance through their employers (even with the same employer), they are not usually linked on record. Likewise, most clinics and hospitals list emergency contact as optional (instead of mandatory) information. Most patients provide one or two emergency contacts, but not their entire family when filling the form – The Polubriaginof study (Polubriaginof et al., 2018) collected on average 1.86 emergency contacts per patient.

To address these issues, we propose a new data source (online obituaries) and a special Natural Language Processing (NLP) solution for systematically constructing biological relationships for large families of multi-generations. Obituaries contain rich and high-quality kinship information and are publicly available from the sites of newspapers and funeral services companies. Although obituaries are similar to social media, they are much less studied in biomedicine. One study analyzed obituaries to investigate cancer mortality trends (Tourassi et al., 2016). Another group combined LinkedIn profiles and obituaries to investigate the association between frequent relocation and lung cancer risk (Yoon et al., 2015). In this project, our ultimate goal is to link multiple obituaries by cross-validating name, age, residence and birth/death date information, to build large family trees. For this paper, we aim to investigate if state-of-the-art NLP methods can automatically extract names and kinships from online obituaries with high accuracy.

Establishing human names and their relations is a Named Entity Recognition (NER) and Relation Extract (RE) task. The NLP community has been working on both for many years. Usually, NER and RE are considered as two separate and sequential tasks (NER precedes RE). Most information extraction systems in biomedicine, including those mining biomedical literature to extract adverse drug events, and molecular interactions between drug, gene and proteins, are built on a battery of pipeline modules integrating NER and RE tasks (Miwa et al., 2012; Kang et al., 2014; Yildirim et al., 2014; Sun et al., 2017; Li et al., 2013; Li et al., 2017). However, pipeline models have inherent limitations: (i) The error from NER will propagate to RE. (ii) Pipeline models cannot fully utilize the internal connections between NER and RE to improve model performance when the separated models finished the two tasks independently. For instance, in a task of extracting adverse drug event, the named entity appeared before the relation keyword of "induce" (non-passive voice) would be a drug and the named entity after "induce" would be an adverse event. NER, which should be finished firstly, definitely would be harder to benefit from this relation information than RE. (iii) Pipeline models are computationally redundant and error-prone because they match up every two named entities to decide their relations, which is not necessary.

In this work, we propose a joint neural model to simultaneously extract names and kinships from obituaries, which combines a two-layer bi-directional Long Short-Term Memory (bi-LSTM) (Hochreiter and Schmidhuber, 1997) and a unique tagging scheme. It, in theory, surpasses pipeline models by overcoming the limitations (i) (Li et al., 2016; Zheng et al., 2017a) and (iii), and by making room for leveraging the contextual information and domain knowledge to address limitation (ii). The rest of the paper is organized into four sections. In the Data and Methods section, we describe how we annotated the obituary corpus, together with the special tagging scheme, the bi-directional LSTM model and evaluation metrics. Then in the Results section, we demonstrate corpus statistics and model performance metrics. After that, we share some discussions regarding the strengths and limitations of our method, before final conclusions and future work.

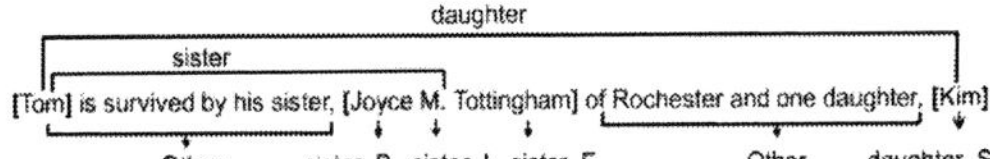

Figure 1: A novel tagging scheme for extracting names and kinships from obituaries

2 Data and Methods

2.1 Corpus preparation

We downloaded obituaries from the websites of three funeral services and one local newspaper in Rochester Minnesota, including: (1) http://www.bradshawfuneral.com, (2) http://www.czaplewskifuneralhomes.com, (3) https://mackenfuneralhome.com, and (4) https://www.postbulletin.com. The downloaded obituaries were published from 10/2008 to 09/2018. After removing those shorter than 290 characters, which is unlikely to contain any mentions of family relationships, messy ones with irregular HTML format or language, and duplicates, we selected 1,809 obituaries for annotation, due to limited resources and labor-intensive annotation described in next subsection.

2.2 Corpus annotation

The success of a machine learning application does not solely depend on the model itself. Most of the time it is more determined by the quality of data, particularly the gold standard dataset for training and testing the model. The challenge for annotating a natural language corpus is that the ground truth is not always obvious, due to the ambiguity and complexity of human language. A detailed annotation guideline and duplicated annotation by multiple people is often necessary to guarantee annotation consistency and corpus quality. Based on two examples of biomedical corpus annotations (Gurulingappa et al., 2012, Roberts et al., 2009), we designed an iterative annotation workflow and revised our guideline three times. All annotations were done at the document level so that the annotators can leverage the context in difficult cases. An open-source software called MAE version 2.2.6 (Kyeongmin, 2016) was used as the annotation tool throughout the entire process.

The corresponding author and three native speakers of English drafted the 1st version of annotation guideline. Then 3 computer science major students were trained for annotation in 2

	Precision (%)	Recall (%)	F1 score (%)
Training round 1	67.93	69.54	62.21
- Last name distribution	70.93	73.63	65.58
- Name with parenthesis	72.35	73.16	66.06
- Name-Residence Pair	69.32	71.11	63.67
- All features	76.84	78.98	71.01
Training round 2	74.61	76.31	68.92
- Last name distribution	77.71	80.51	72.40
- Name with parenthesis	79.03	79.94	72.77
- Name-Residence Pair	76.03	77.94	70.43
- All features	83.66	85.91	77.87
Final annotation	88.46	88.58	82.80
- Last name distribution	89.86	89.96	84.19
- Name with parenthesis	89.26	89.43	83.62
- Name-Residence Pair	88.58	88.68	82.91
- All features	90.94	91.05	85.27

Table 1: IAA scores in different rounds of annotation with different annotation features (- means "without")

rounds. In each round, we randomly selected 300 obituaries and asked each student to annotate 200 obituaries. This way each obituary was annotated twice by two different annotators. At the end of each round of training, we evaluated the annotation consistency using inter-annotator agreement (IAA) metrics and improved the annotation guideline. Considering that extracting kinship was actually a NER+RE task, we adopted precision, recall and F1 score rather than Kappa coefficient to report IAA, as suggested by Gurulingappa et al., 2012 and Chinchor, 1992.

After completing the training, 3 qualified annotators finished annotating the rest obituaries with the assistance of a rule-based quality control program written by us. Table 1 demonstrates that the precision, recall and F1 score were steadily improving through training round 1, training round 2 and final annotation. The discrepancy in the final annotation was resolved through group discussions. We warranted that 1,809 obituaries have high-quality annotations before building the models.

2.3 The tagging scheme

Conventional NER and RE are usually formulated as triplet tagging (entity_1, relation, entity_2). But our addressed task is not a general NER+RE task. It is simplified by three factors: (1) There is only one type of named entity to detect (human names); (2) all relations have the same first entity (the deceased); and (3) the first entity is mentioned in the metadata or the first sentence of the obituary, and hence does not need to be detected most of the times. Therefore in this study, we proposed a novel tagging scheme inspired by Zheng et al (Zheng et. al, 2017b), which extracts names and kinships in relative to the deceased person in one step, as shown in Figure 1. We used the popular "BIESO" (begin, inside, end, single, and other) scheme to mark the position of words in entities, where "O" refers to cases that a word does not belong to an entity. This way we can identify a named entity by simply applying the rule of S or B + n*I + E, where n ≥ 0. But we added the kinship type into the "BIESO" tags, in order to synchronize the NER and RE annotation. So each tag consists of two parts: the first part indicates the kinship type and the second part illustrates the position of a word in an entity. In an illustrative example shown in Figure 1, "Joyce M. Tottingham" is assigned three tags, including "sister_B" for the word "Joyce", "sister_I" for the word "M.", and "sister_E" for the word "Tottingham". For the single-word entity "Kim", the assigned tag was "daughter_S". All the remaining words were assigned a tag "O". Because we set the decreased as the default first entity for any kinship, triplets were simplified to duplets, like [sister, Joyce M. Tottingham] and [daughter, Kim] for the sentence in Figure 1. "Tom" was the name of the deceased person (inferred from the context or metadata) and we did not annotate it as a named entity. But we annotated other entity types including age, residence, birth date and death date. We plan to use these additional entity types in future work when we build the family trees and link them to EHR database.

2.4 The end-to-end joint neural model

The end-to-end neural model has lately demonstrated effectiveness in various NLP tasks, including NER, RE, part-of-speech tagging and semantic role labeling (Hashimoto et al., 2017, Strubell, 2018). In this study, we adopted an end-to-end neural model (See Figure 2), which contained an embedding layer, two bi-LSTM layers, and a softmax output layer. A rule-based result improver layer was also added to the end for consolidating the tags generated by the softmax output layer. We also used a dynamically weighted loss function to alleviate data imbalance issue.

The input sentences were tokenized and each token was converted to a word vector learned

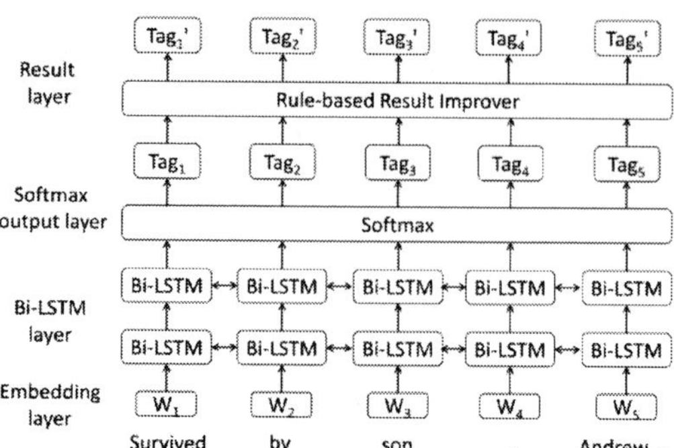

Figure 2: The neural network architecture for jointly extracting names and kinship types

from the GloVe method (Pennington et al., 2014), when fed into the embedding layer. Padding, which was a common programming trick, was performed in a way that all sentences were aligned to the longest sentence in a batch using padding tags for parallel computation. They would not impact the model performance as the output of those padding tags were masked out in the backward layer of the Bi-LSTM model. The Bi-LSTM architecture consisted of a forward layer and a backward layer, which was supposed to capture sequential context information bi-directionally. Both layers consisted of blocks made up of a forget gate, an input gate and an output gate. The forget gate decided how much information from the previous block would be dropped at the current block, considering the current input and the previous hidden representation. The input gate took the output of the forget gate and the previous cell state to update the current cell state. The output gate was designed to create a hidden representation for each token based on all the information from the forget gate and input gate. Finally, the outputs of both forward layer and backward layer were concatenated by Bi-LSTM as final representation. The softmax function served as the classifier for computing final normalized probabilities for each tag. After that, each token was classified into one of (m*5+1) tags, where m was the total number of kinship types. We tried m=57 and 34, according to the number of annotated examples in our experiment (See Table 4). In the end, a rule-based result improver was added to make sense of the sequence of the classified tags. For example, if the softmax output layer tagged two neighboring words as "sister_B" and "sister_I" without "sister_E" nearby, the improver would correct the second tag to "sister_E".

Corpus	Count	Deceased Person	Count	Special Language Patterns	Count
Sentences	30,035	Name	1,711	Last name distribution	5,186
Names	29,938	Mention of Age	1,517	Name with parentheses	8,118
Kinship	27,227	Mention of Death Date	1,712	• Nickname	84
Mention of Residence	8,476	Mention of Birth Date	1,522	• Previous last name	1,607
Name-Residence Pair	9,189	Mention of Residence	1,331	• Spouse's name	6,427

Table 2: Summary Statistics of the Corpus

Language Pattern		Example	Explanation
Last name distribution		Preceded in death by her grandparents, Ellen and Everett Uebel.	Uebel is also the last name for Ellen.
Name with parenthesis	Nickname	Kay is also survived by her daughter Maureen (Mo) Bahr of Rochester	Mo is the nickname of Maureen Bahr.
	Previous last name	Paul was born April 18, 1942 in Rochester to Boyd and Fern (Miller) Kinyon.	Miller is the maiden name for Fern Kinyon.
	Spouse's name	Survived by daughter, Sydney (Sam) Davis; granddaughter, Autumn Ellen.	Sydney Davis's husband is Sam Davis.

Table 3: Examples of unique language patterns in obituaries

Hierarchy	Kinship type
Generation 0	ex-husband (18), ex-wife (32), married to (1,457), spouse (18), husband (586), wife (690), sibling (718), cousin (91), brother (2,106), sister (2,156), half-brother (13), half-sister (7), sister-in-law (344), sibling-in-law (28), cousin-in-law (1), brother-in-law (251)
Generation 1	child (2,658), daughter (1,445), son (1,713), niece (242), nephew (297), step-child (175), step-daughter (60), step-son (65), child-in-law (25), daughter-in-law (114), son-in-law (103), niece-in-law (20), nephew-in-law (25)
Generation 2	grandson (310), grandchild (4,413), granddaughter (231), grandnephew (24), grandniece (24), grandson-in-law (13), grandchild-in-law (11), granddaughter-in-law (12), step-grandchild (98), step-grandson (7), step-granddaughter (6)
Generation 3	great grand-child (1,293), great granddaughter (46), great grandson (65), great grand-nephew (2), great grand-niece (6), great grandchild-in-law (4),
Generation 4	great-great grand-child (27), great-great granddaughter (1), great-great grandson (1)
Generation -1	born to (2,332), son of (132), daughter of (172), parent (720), mother (155), father (139), step-mother (16), step-father (24), step-parent (2), aunt (49), uncle (54), parent-in-law (43), mother-in-law (30), father-in-law (26), aunt-in-law (6), uncle-in-law (3)
Generation -2	grandparent (210), grandmother (44), grandfather (29), grand uncle (1), grandmother-in-law (1)
–	Other* (987)

Table 4: 71 kinship types in annotated obituaries. Top 5 common relationships are highlighted in red.
* Other relationships refer to kinships not included in previous 6 categories, such as fiancé, guardian, and friend.

Dynamic weighted Loss function: We trained our joint model with weighted log-likelihood function, and used RMSprop (Tieleman and Hinton, 2012) for optimization. The objective function was defined as follows:

$$L = -\sum_{s=1}^{B}\sum_{t=1}^{L_s}(\log(p_t^{(s)} = \hat{y}_t^{(s)}|x_s) \cdot (1 - P(O))$$
$$+ f_\omega \cdot \log(p_t^{(s)} = \hat{y}_t^{(s)}|x_s)$$
$$\cdot P(O) + \frac{\lambda}{2}\|\theta\|_2^2 \tag{1}$$

Kinship filter	Method	Average method	Precision (%)	Recall (%)	F-measure (%)
n≥10	Pipeline	macro	68.60 (4.81)	69.52 (4.98)	68.43 (4.90)
		micro	87.10 (0.57)	89.46 (0.82)	87.80 (0.78)
	Joint	macro	72.69 (3.96)	78.54 (3.85)	74.93 (3.95)
		micro	95.74 (0.98)	98.25 (0.43)	96.98 (0.60)
n≥50	Pipeline	macro	81.11 (3.70)	79.51 (2.62)	79.18 (3.22)
		micro	85.42 (0.98)	92.80 (0.43)	88.18 (0.60)
	Joint	macro	85.27 (3.90)	94.35 (2.09)	88.97 (3.18)
		micro	96.06 (0.64)	98.12 (0.37)	97.08 (0.46)

Table 5: Comparing the performance of pipeline model versus joint model. The values in brackets represent the standard deviation during 10-fold cross validation.

where B was the batch size, L_s was the length of input sentence x_s. $\hat{y}_t^{(s)}$ and $p_t^{(s)}$ were the true tag and the normalized probability of the predicted tag for word t. λ was the hyper-parameter for L2 regularization. $P(O)$ was the indicator function to determine if the current tag was "O" (other), which was formulated as:

$$P(O) = \begin{cases} 0, & if\ tag = "O" \\ 1, & if\ tag \neq "O" \end{cases} \quad (2)$$

f_ω was dynamic weighted loss function, which assigned the tag ω different weights in different sentences, aiming to alleviate influence caused by too much "O" tag. It was defined as:

$$f_\omega = \frac{\frac{\sum_{j \in T} N_{D_i}^j}{N_{D_i}^\omega} - N_{Y min}}{N_{Y max} - N_{Y min}} \quad (3)$$

where T was the union of all possible tags, D_i referred to a sentence i in a batch of the training set, $N_{D_i}^\omega$ was the total count of all tags in D_i, $N_{D_i}^j$ was the number of a specific tag ω in D_i, and $N_{Y max}$ and $N_{Y min}$ were the maximal and minimal hyper-parameters for normalization respectively.

2.5 Evaluation metrics

A recognized named entity mention was considered true positive (TP) if both its boundary and type matched with the annotation. A relation extraction was considered as TP if both the NER and RE tasks were correctly captured. A recognized entity or relation was considered as false positive (FP) if it did not exactly match with the manual annotation in terms of the boundaries and relation types. The number of false negatives (FN) instances was computed by counting the number of named entities or relations in the manual annotation that had been missed by the model.

We performed 10-fold cross validation in our experiment, where 10% of the annotated data were randomly selected for validation, and the remaining for training the model. We evaluated the model performance using macro- and micro-averaged Precision, Recall and F-measure. A macro-averaged metric treats all classes equally by computing the metric independently for each class and then taking the average. In contrast, a micro-averaged metric aggregates the TP, TN, FP, and FN counts of all classes to compute an average metric.

Our corpus and codes could be downloaded at https://github.com/qw52025804/Obituary.git.

3 Results

3.1 Corpus annotation

Table 2 lists the detailed summary statistics of our corpus. There were 1,711 mentions of deceased names in 1,809 obituaries. Some obituaries mentioned the names of the deceased people in the title (metadata) rather than the main body of obituaries. In those cases, we directly linked the deceased names in the title of obituaries with their main body of free text. On average, each obituary

<table>
<tr><th colspan="2" align="center">Examples of Correct Classification</th></tr>
<tr><th align="center">Sentence</th><th align="center">Extracted Relation</th></tr>
<tr><td>On May 8, 1982 he married Madonna Oleson & became a proud dad of Ryan and Kelly.</td><td>Madonna Oleson : wife
Ryan : child
Kelly : child</td></tr>
<tr><td>He is survived by his brother Richard R. Arend (Carol) of Rochester, his beloved children and their mother, Kristy.</td><td>Richard R. Arend (Carol): brother
Kristy : wife</td></tr>
<tr><td>One brother, Gordon "Scotty" Hyland of LaMirada, CA. and many nieces and nephews.</td><td>Gordon "Scotty" Hyland : brother</td></tr>
<tr><th colspan="2" align="center">Examples of Wrong Classification</th></tr>
<tr><th align="center">Sentence</th><th align="center">Extracted Relation</th></tr>
<tr><td>Craig is also survived by the boy's mother, Jolene Stock, sister Dianna Povilus; ...</td><td>Jolene Stock : mother</td></tr>
<tr><td>Survivors include Mary, his wife of 44 years and three children. Kristen (Matt) Asleson of Fountain, MN, and ...</td><td>Kristen (Matt) Asleson : grand child</td></tr>
<tr><td>Wooing Cecelia Stevens by serenading the words from the musical Carousel, "If I loved you, words wouldn't come in an easy way" - he proposed and on July 6, 1955, they began sixty-one years of marriage.</td><td>Cecelia Stevens : missing</td></tr>
</table>

Table 6: Correctly classified examples and wrongly classified examples

contains 16.6 sentences, or 1,809 obituaries contain 30,035 sentences in total. We extracted and annotated 29,938 names, 27,227 family relations and 8,476 residences for the deceased and their families. We were able to pair up a name and a residence for 9,189 times. For the deceased people, we also annotated their age, death date, birth date, and residence when available.

We noticed two interesting language patterns in obituaries, namely last name distribution and name with parentheses (See Table 3). These patterns might be due to the word limitation in the old time when the family paid for publishing an obituary on printed newspapers. In total, we annotated 71 kinships (See Table 4). Among them, 57 kinships have ≥10 examples, 34 kinships have ≥50 examples, and 28 relationships have ≥ 100 examples. The most populated five relationships were grandchild (4,413), child (2,658), born to (equivalent to parent, 2,332), sister (2,156) and brother (2,106).

It is worth noting that we kept "married to" and "spouse", "born to" and "parent" as separate kinship types in our experiment. This is because the syntax, co-occurred words and their order near "married to"/"born to" are subtly different from "spouse"/"parent". Keeping them as separate kinship types might help to improve the model performance. We will group them in the next step when we build the family trees, as they are semantically equivalent.

3.2 Model performance

Table 5 illustrates the final performance of the baseline method (pipeline model) versus our proposed joint neural model for extracting names and kinships from obituaries. The baseline model consists of two one-layer bi-LSTMs. The first bi-LSTM is for NER with simple BIESO tagging scheme, and its outputs were used as the inputs of the second bi-LSTM for RE. The general architecture is the same as that of the joint model, but the tagging scheme is different for NER, and NER and RE worked in a pipelined way. It is shown that the joint model outperformed the pipeline model by 4.09%, 9.02% and 6.5% for Precision, Recall and F measure at macro level using 57 kinships with 10 or more examples. The joint model outperformed the pipeline model by even bigger margins for Precision, Recall and F measure (4.16%, 14.84% and 9.79% respectively) at macro level when considering 34 kinships with 50 or more examples. The micro-level evaluation metrics demonstrated even better results of similar trends, due to the nature of an imbalanced multi-class classification problem. Table 6 showed some correctly classified examples and wrongly classified examples, which demonstrated the challenges in this project.

4 Discussions

The proposed joint neural model seemed capable of extracting the human names and relations with high performance. For common kinship types with large number of examples in the training dataset, such as grandchild, child, parent (born to), sister and brother, the model's performance were close to perfect: Precision> 96.06%, Recall>98.12% and F measure> 97.08%. It could also recognize multiple variations of family relationships such as "marry" and "dad of", thanks to the high quality annotated corpus we created.

As shown in Table 6, the model was able to tell that "Kristy" was the wife of the deceased person (the second example of correct classification), but could not figure out "Jolene Stock" was the wife of the deceased "Craig" (the first example of wrong classification). It seems that the model was confused by the relationships between the deceased, "the boy's mother" and Jolene Stock. For the second example of wrong classification, the incorrect punctuation might have led to the error. The period before "Kristen (Matt) Asleson" should be a comma instead. The last example in Table 6 was an extremely difficult and rare case. Common kinship keyword indicating wife was missing. Without properly understanding the semantic meaning of 'propose' and 'marriage' in the sentence, our model failed to pick up "Cecelia Stevens" as a name.

One limitation of this study was that we built the Bi-LSTM model on sentences, and therefore lost the context information beyond a sentence. More sophisticated LSTM model would be helpful to parse the entire document of obituaries. Another challenge was that we could not afford to annotate more obituaries, which led to 14 kinship types had less than 10 examples (e.g., grandmother-in-law, grand uncle, great-great grandson and great-great granddaughter). Our model, or any supervised models, would not perform well on such small size of training data.

5 Conclusions and Future Work

In this work, we built an annotated corpus of >30,000 sentences (from 1,809 obituaries written in English) and proposed a two-layer Bi-LSTM model to simultaneously extract human names and kinships. Our joint neural model achieved macro-averaged Precision, Recall and F measure of 72.69%, 78.54% and 74.93%, and micro-averaged Precision, Recall and F measure of 95.74%, 98.25% and 96.98% using 57 kinships with 10 or more examples during 10-fold cross validation experiment. The model performance improved dramatically when trained with 34 kinships with 50 or more examples. We shared our corpus and codes on GitHub for the convenience of researchers.

Given such promising results, we will continue to improve our joint model to recognize other types of entity and relation, including the age, residence, birth date and death date. We will further parse names with parenthesis; resolve last name distributions; and leverage existing knowledge to infer the gender of names. Only when we complete theses tasks with high quality, could we build large family trees and link people to our EHR database. We are cautiously optimistic because almost all residents in Rochester MN have been patients at Mayo Clinic at some time of their life and population mobility rate in Rochester MN is far less than major metropolitan areas in the U.S. With the massive obituary data freely available on the Internet, our ultimate goal is to accelerate large-scale disease heritability research and clinical genetics research.

6 Ethics

In this study, we mined only publicly available information from 4 websites, without interacting with, intervening, or manipulating/changing the website's environment. The study does not include "human subject" data and is approved by the Office of Research and Compliance without IRB requirement at Mayo Clinic.

7 Acknowledgements

Funding for KH, JW, XM, CZ and CL are provided by the National Key Research and Development Program of China (2018YFC0910404); National Natural Science Foundation of China (61772409) and the consulting research project of the Chinese Academy of Engineering (The Online and Offline Mixed Educational Service System for "The Belt and Road" Training in MOOC China). Funding for MH and LY are provided by the National Center for Advancing Translational Sciences (UL1TR002377) and the National Library of Medicine (5K01LM012102).

8 References

Alvaro, N., Miyao, Y., Collier, N.: TwiMed: Twitter and PubMed Comparable Corpus of Drugs, Diseases, Symptoms, and Their Relations. *JMIR*

Public Health and Surveillance (2017). doi:10.2196/publichealth.6396

Bekoulis, G., Deleu, J., Demeester, T., Develder, C.: Adversarial training for multi-context joint entity and relation extraction, 2830{2836 (2018). 1808.06876

Chatterjee, N., Shi, J., García-Closas, M.: Developing and evaluating polygenic risk prediction models for stratified disease prevention. *Nature Reviews Genetics* 17(7), 392{406 (2016). doi:10.1038/nrg.2016.27

Chinchor, N.: MUC-4 evaluation metrics. In: Proceedings of the 4th Conference on Message Understanding MUC4 '92 (1992). doi:10.3115/1072064.1072067. arXiv:1011.1669v3

Cohen, K.B., Fox, L., Library, D., Ogren, P.V., Hunter, L.: Corpus design for biomedical natural language processing. Technical report (2005). http://compbio.uchsc.edu/corpora

Cunningham, H., Maynard, D., Bontcheva, K., Tablan, V.: GATE: A framework and graphical development environment for robust NLP tools and applications. In: *Proceedings of the 40th Anniversary Meeting of the Association for Computational Linguistics*, ACL 2002 (2002). doi:10.3115/1073083.1073112

Denny J C, Ritchie M D, Basford M A, et al. PheWAS: demonstrating the feasibility of a phenome-wide scan to discover gene–disease associations. *Bioinformatics*, 2010, 26(9): 1205-1210.

Ge, T., Chen, C.-Y., Neale, B.M., Sabuncu, M.R., Smoller, J.W.: Phenome-wide heritability analysis of the UK Biobank. *PLOS Genetics* 13(4), 1006711(2017). doi:10.1371/journal.pgen.1006711

Ginn, R., Pimpalkhute, P., Nikfarjam, A., Patki, A., O 'connor, K., Sarker, A., Smith, K., Gonzalez, G.: Mining Twitter for Adverse Drug Reaction Mentions: A Corpus and Classification Benchmark. In: *Proceedings of the Fourth Workshop on Building and Evaluating Resources for Health and Biomedical Text Processing* (2014). doi:10.1590/S1516-35982012000500024

Gurulingappa, H., Rajput, A.M., Roberts, A., Fluck, J., Hofmann-Apitius, M., Toldo, L.: Development of a benchmark corpus to support the automatic extraction of drug-related adverse effects from medical case reports. *Journal of Biomedical Informatics* 45(5), 885{892 (2012). doi:10.1016/j.jbi.2012.04.008

Hashimoto, Kazuma, Yoshimasa Tsuruoka, and Richard Socher. "A Joint Many-Task Model: Growing a Neural Network for Multiple NLP Tasks." *Proceedings of the 2017 Conference on Empirical Methods in Natural Language Processing*. 2017.

Herrero-Zazo, M., Segura-Bedmar, I., Mart´ınez, P.: Annotation Issues in Pharmacological Texts. Procedia-Social and Behavioral Sciences 95, 211{219 (2013). doi:10.1016/j.sbspro.2013.10.641

Hochreiter, S., Schmidhuber, J.: Long short-term memory. *Neural computation* 9(8), 1735{1780 (1997)

Kang, N., Singh, B., Bui, C., Afzal, Z., van Mulligen, E.M., Kors, J.A.: Knowledge-based extraction of adverse drug events from biomedical text. *BMC Bioinformatics* 15(1) (2014). doi:10.1186/1471-2105-15-64

Kyeongmin Rim: MAE2: Portable Annotation Tool for General Natural Language Use (May), 75{80 (2016)

Li, F., Yue, Z., Meishan, Z., Ji, D.: Joint models for extracting adverse drug events from biomedical text. *International Joint Conference on Artificial Intelligence* 2016-Janua, 2838{2844 (2016)

Li, C., Liakata, M., Rebholz-Schuhmann, D.: Biological network extraction from scientific literature: State of the art and challenges. *Briefings in Bioinformatics* 15(5), 856-877 (2013). doi:10.1093/bib/bbt006

Li, F., Zhang, M., Fu, G., Ji, D.: A neural joint model for entity and relation extraction from biomedical text. *BMC Bioinformatics* 18(1), 1-11 (2017). doi:10.1186/s12859-017-1609-9

MacKinlay, A., Aamer, H., Yepes, A.J.: Detection of Adverse Drug Reactions using Medical Named Entities on Twitter. *AMIA Annual Symposium Proceedings* 2017, 1215{1224 (2017)

Mayer, J., Kitchner, T., Ye, Z., Zhou, Z., He, M., Schrodi, S.J., Hebbring, S.J.: Use of an Electronic Medical Record to Create the Marshfield Clinic Twin/Multiple Birth Cohort. *Genetic Epidemiology* 38(8), 692-698 (2014). doi:10.1002/gepi.21855

Miwa, M., Bansal, M.: End-to-End Relation Extraction using LSTMs on Sequences and Tree Structures (2016). doi:10.18653/v1/P16-1105. 1601.00770

Miwa, M., Thompson, P., McNaught, J., Kell, D.B., Ananiadou, S.: Extracting semantically enriched events from biomedical literature. *BMC Bioinformatics* (2012). doi:10.1186/1471-2105-13-108

Pennington, J., Socher, R., Manning, C.: Glove: Global vectors for word representation. In: *Proceedings of the 2014 Conference on Empirical Methods in Natural Language Processing*, pp. 1532-1543 (2014)

Polderman, T.J.C., Benyamin, B., de Leeuw, C.A., Sullivan, P.F., van Bochoven, A., Visscher, P.M., Posthuma.D.: Meta-analysis of the heritability of human traits based on fifty years of twin studies. *Nature Genetics* 47(7), 702-709 (2015). doi:10.1038/ng.3285

Polubriaginof, F.C.G., Vanguri, R., Quinnies, K., Belbin, G.M., Yahi, A., Salmasian, H., Lorberbaum, T., Nwankwo, V., Li, L., Shervey, M.M., Glowe, P., Ionita-Laza, I., Simmerling, M., Hripcsak, G., Bakken, S., Goldstein, D., Kiryluk, K., Kenny, E.E., Dudley, J., Vawdrey, D.K., Tatonetti, N.P.: Disease Heritability Inferred from Familial Relationships Reported in Medical Records. *Cell* 173(7), 1692{170411 (2018). doi:10.1016/j.cell.2018.04.032

Roberts, A., Gaizauskas, R., Hepple, M., Demetriou, G., Guo, Y., Roberts, I., Setzer, A.: Building a semantically annotated corpus of clinical texts. *Journal of Biomedical Informatics* 42(5), 950{966 (2009). doi:10.1016/j.jbi.2008.12.013

Robinson P N. Deep phenotyping for precision medicine. *Human mutation*, 2012, 33(5): 777-780.

Strubell, E., Verga, P., Andor, D., Weiss, D., McCallum, A.: Linguistically-informed self-attention for semantic role labeling. arXiv preprint arXiv:1804.08199 (2018)

Strubell, Emma, et al. "Linguistically-Informed Self-Attention for Semantic Role Labeling." *Proceedings of the 2018 Conference on Empirical Methods in Natural Language Processing*. 2018.

Sun, T., Zhou, B., Lai, L., Pei, J.: Sequence-based prediction of protein protein interaction using a deep-learning algorithm. *BMC Bioinformatics* (2017). doi:10.1080/01418639108224439

Tieleman, T., Hinton, G.: Lecture 6.5-rmsprop: Divide the gradient by a running average of its recent magnitude. COURSERA: Neural networks for machine learning 4(2), 26{31 (2012)

Tourassi G, Yoon HJ, Xu S. A novel web informatics approach for automated surveillance of cancer mortality trends. Journal of biomedical informatics. 2016 Jun 1;61:110-8. doi: 10.1016/j.jbi.2016.03.027

Vaswani, A., Shazeer, N., Parmar, N., Uszkoreit, J., Jones, L., Gomez, A.N., Kaiser, L., Polosukhin, I.: Attention is all you need. In: *Advances in Neural Information Processing Systems*, pp. 5998{6008 (2017)

Verga, P., Strubell, E., McCallum, A.: Simultaneously Self-Attending to All Mentions for Full-Abstract Biological Relation Extraction (2018). doi:10.18653/v1/N18-1080. 1802.10569

Wang, K., Gaitsch, H., Poon, H., Cox, N.J., Rzhetsky, A.: Classification of common human diseases derived from shared genetic and environmental determinants. *Nature Genetics* 49(9), 1319{1325 (2017). doi:10.1038/ng.3931

Yildirim, P., Majnari´c, L., Ekmekci, O.I., Holzinger, A.: Knowledge discovery of drug data on the example of adverse reaction prediction. *BMC Bioinformatics* (2014). doi:10.1186/1471-2105-15-S6-S7

Yoon HJ, Tourassi G, Xu S. Residential mobility and lung cancer risk: Data-driven exploration using internet sources. In International Conference on Social Computing, Behavioral-Cultural Modeling, and Prediction 2015 Mar 31 (pp. 464-469). Springer, Cham. doi: 10.1007/978-3-319-16268-3_60

Zheng, S., Hao, Y., Lu, D., Bao, H., Xu, J., Hao, H., Xu, B.: Joint entity and relation extraction based on a hybrid neural network. *Neurocomputing* 257, 59{66 (2017a). doi:10.1016/j.neucom.2016.12.075

Zheng, S., Wang, F., Bao, H., Hao, Y., Zhou, P., Xu, B.: Joint Extraction of Entities and Relations Based on a Novel Tagging Scheme (2017b). doi:10.24963/ijcai.2018/620. 1706.05075

Lexical Normalization of User-Generated Medical Forum Data

Anne Dirkson, Suzan Verberne & Wessel Kraaij
LIACS, Leiden University
Niels Bohrweg 1, Leiden, the Netherlands
{a.r.dirkson, s.verberne, w.kraaij}@liacs.leidenuniv.nl

Abstract

In the medical domain, user-generated social media text is increasingly used as a valuable complementary knowledge source to scientific medical literature. The extraction of this knowledge is complicated by colloquial language use and misspellings. Yet, lexical normalization of such data has not been addressed properly. This paper presents an unsupervised, data-driven spelling correction module for medical social media. Our method outperforms state-of-the-art spelling correction and can detect mistakes with an $F_{0.5}$ of 0.888. Additionally, we present a novel corpus for spelling mistake detection and correction on a medical patient forum.

1 Introduction

In recent years, user-generated data from social media that contains information about health, such as patient forum posts or health-related tweets, has been used extensively for medical text mining and information retrieval (IR) (Gonzalez-Hernandez et al., 2017). This user-generated data encapsulates a vast amount of knowledge, which has been used for a range of health-related applications, such as the tracking of public health trends (Sarker et al., 2016) and the detection of adverse drug responses (Sarker et al., 2015). However, the extraction of this knowledge is complicated by non-standard and colloquial language use, typographical errors, phonetic substitutions, and misspellings (Clark and Araki, 2011; Sarker, 2017; Park et al., 2015). Thus, social media text is generally noisy and this is only aggravated by the complex medical domain (Gonzalez-Hernandez et al., 2017).

Despite these challenges, text normalization for medical social media has not been explored thoroughly. Medical lexical normalization methods (i.e. abbreviation expansion (Mowery et al., 2016) and spelling correction (Lai et al., 2015; Patrick et al., 2010)) have mostly been developed for clinical records or notes, as these also contain an abundance of domain-specific abbreviations and misspellings. However, social media text presents distinct challenges, such as colloquial language use, (Gonzalez-Hernandez et al., 2017; Sarker, 2017) that cannot be tackled with these methods.

The most comprehensive benchmark for general-domain social media text normalization is the ACL W-NUT 2015 shared task[1] (Baldwin et al., 2015). The current state-of-the-art system for this task is a modular pipeline with a hybrid approach to spelling, developed by Sarker (2017). Their pipeline also includes a customizable back-end module for domain-specific normalization. However, this back-end module relies, on the one hand, on a standard dictionary supplemented manually with domain-specific terms to detect mistakes and, on the other hand, on a language model of generic Twitter data to correct these mistakes. For domains that have many out-of-vocabulary (OOV) terms compared to the available dictionaries and language models, such as medical social media, this is problematic.

Manual creation of specialized dictionaries is an unfeasible alternative: medical social media can be devoted to a wide range of different medical conditions and developing dictionaries for each condition (including laymen terms) would be very labor-intensive. Additionally, there are many different ways of expressing the same information and the language use in the forum evolves over time. Consequently, hand-made lexicons may get outdated (Gonzalez-Hernandez et al., 2017). In this paper, we present an alternative: a corpus-driven spelling correction approach. We address two research questions:

1. To what extent can corpus-driven spelling correction reduce the out-of-vocabulary rate in medical social media text?

---

[1] https://noisy-text.github.io/norm-shared-task.html

Proceedings of the 4th Social Media Mining for Health Applications (#SMM4H) Workshop & Shared Task, pages 11–20
Florence, Italy, August 2, 2019. ©2019 Association for Computational Linguistics

2. To what extent can our corpus-driven spelling correction improve accuracy of health-related classification tasks with social media text?

Our contributions are (1) an unsupervised data-driven spelling correction method that works well on specialized domains with many OOV terms without the need for a specialized dictionary and (2) the first corpus for evaluating mistake detection and correction in a medical patient forum.[2]

Our method is designed to be conservative and to focus on precision to mitigate one of the major challenges of correcting errors in domain-specific data: the loss of information due to the 'correction' of already correct domain-specific terms. We hypothesize that a dictionary-based method is able to retrieve more mistakes than a data-driven method, because all terms *not* included in the dictionary are classified as mistakes, which will probably include all non-word errors. However, we also expect that a dictionary-based method will misclassify more correct terms as mistakes, because any domain-specific terms not present in the dictionary will be classified incorrectly.

2 Related work

Challenges in correcting spelling errors in medical social media A major challenge for correcting spelling errors in small and highly specialized domains is a lack of domain-specific resources. This complicates the automatic creation of relevant dictionaries and language models. Moreover, if the dictionaries or language models are not domain-specific enough, there is a high probability that specialized terms will be incorrectly marked as mistakes. Consequently, essential information may be lost as these terms are often key to knowledge extraction tasks (e.g. a drug name) and to specialized classification tasks (e.g. does the post contain a side effect of drug X?).

This challenge is further complicated by the dynamic nature of language on medical social media: in both the medical domain and social media novel terms (e.g. a novel drug names) and neologisms (e.g. group-specific slang) are constantly introduced. Unfortunately, professional clinical lexicons are also unsuited for capturing the domain-specific terminology on forums, because laypersons and health care professionals express health-related concepts differently (Zeng and Tse, 2006).

Another complication is the frequent misspellings of key medical terms, as medical terms are typically difficult to spell (Zhou et al., 2015). This results in an abundance of common mistakes in key terms, and thus, a large amount of lost information if these terms are not handled correctly.

Lexical normalization of generic social media In earlier research, text normalization for social media was mostly unsupervised or semi-supervised e.g. (Han et al., 2012) due to a lack of annotated data. These methods often pre-selected and ranked correction candidates based on phonetic or lexical string similarity (Han et al., 2012, 2013). Han et al. (2013) additionally used a tri-gram language model trained a large Twitter corpus to improve correction. Although these methods did not rely on training data to correct mistakes, they did rely on dictionaries to determine whether a word *needed* to be corrected (Han et al., 2012, 2013). The opposite is true for modern supervised methods, which rely on training data but not on dictionaries. For instance, the best performing method at the ACL W-NUT shared task of 2015 used canonical forms in the training data to develop their own normalization dictionary (Jin, 2015). The second and third best performing methods were also supervised and used deep learning to detect and correct mistakes (Leeman-Munk et al., 2015; Min and Mott, 2015) (for more detail on W-NUT systems see Baldwin et al. (2015)). Since specialized resources (appropriate dictionaries or training data) are not available for medical forum data, a method that relies on neither is necessary. We address this gap.

Additionally, recent approaches often make use of language models, which require a large corpus of comparable text from the same genre and domain (Sarker, 2017). This is however a major obstacle for employing such an approach in niche domains. Since forums are often highly specialized, the resources that could capture the same language use are limited. Nevertheless, if comparable corpora are available, language models can contribute to effectively reducing spelling errors in social media (Sarker, 2017) due to their ability to capture the context of words and to handle the dynamic nature of language.

3 Data

Medical forum data For evaluating spelling correction methods, we use an international pa-

---

	GIST forum	Reddit forum
# Tokens	1,255,741	4,520,074
# Posts	36,277	274,532
Median post length (IQR)	20 (35)	11 (18)

Table 1: Raw data without punctuation. IQR: Interquartile range

tient forum for patients with Gastrointestinal Stromal Tumor (GIST). It is moderated by GIST Support International (GSI). This data set was donated to us by GSI in 2015. We use a second cancer-related forum to assess generalisability of our methods: a sub-reddit community on cancer, dating from 16/09/2009 until 02/07/2018.[3] It was scraped using the Pushshift Reddit API.[4] The data was collected by looping over the timestamps in the data. This second forum is around 4x larger than the first in terms of tokens (See Table 1).

Annotated data Spelling mistakes were annotated for 500 randomly selected posts from the GIST data. Real word errors and split or concatenation errors were not included, because we are not interested in syntactic or semantic errors (Kukich, 1992). In addition, we considered each word independent of its content, because word bigrams or trigrams are sparse in the small forum collections (Verberne, 2002). Each token was classified as a mistake (1) or not (0) by the first author. A second annotator checked if any of the mistakes were false positives. 53 unique mistakes were found: Their corrections were annotated individually by two annotators. Annotators were provided with the complete post in order to determine the correct word. The initial absolute agreement was 89.0%. If a consensus could not be reached, a third assessor was used to resolve the matter. These 53 mistakes and their corrections form the test set for evaluating spelling correction methods.[5] As far as we are aware, no other spelling error corpora for this domain are publicly available.

In order to tune various thresholds for the detection of spelling mistakes, we split these 500 posts into two sets of 250 posts: a development and a test set. The development set contained 23 mistakes supplemented with a tenfold of randomly selected correct words (230) with the same word length distribution. The development set

was split in a stratified manner into 10 folds for cross-validation. The test set contained 32 unique non-word errors [6], equal to 0.37% of the tokens, supplemented with a tenfold of randomly selected correct words with the same word length distribution.[7]

Spelling error frequency corpus Since by default all edits are weighted equally when calculating Levenshtein distance, we needed to compute a weighted edit matrix in order to assign lower costs and thereby higher probabilities to edits that occur more frequently in the real world. We based our weighted edit matrix on a corpus of frequencies for 1-edit spelling errors compiled by Peter Norvig.[8] This corpus is compiled from four sources: (1) a list of misspellings made by Wikipedia editors, (2) the Birkbeck spelling corpus, (3) the Holbrook corpus and (4) the ASPELL corpus.

Specialized vocabulary for cancer forums To be able to calculate the number of out-of-vocabulary terms in two cancer forums, a specialized vocabulary was created by merging the standard English lexicon CELEX (Burnage et al., 1990) (73,452 tokens), the NCI Dictionary of Cancer Terms (National Cancer Institute) (6,038 tokens), the generic and commercial drug names from the RxNorm (National Library of Medicine (US)) (3,837 tokens), the ADR lexicon used by Nikfarjam et al. (2015) (30,846 tokens) and our in-house domain-specific abbreviation expansions (DSAE) (42 tokens) (see Preprocessing for more detail). As many terms overlapped with those in CELEX, the total vocabulary consisted of 118,052 tokens (62.2% CELEX, 5.1% NCI, 26.1% ADR, 6.5% RxNorm and <0.01% DSAE).

Data sets for external validation We obtained six public classification data sets that use health-related social media data. They were retrieved from the data repository of Dredze[9] and the shared tasks of Social Media Mining 4 Health workshop (SMM4H) 2019[10]. The data sets sizes range from 588 to 16,141 posts (see Table 2).

---

[6]Two errors overlapped between the sets

[7]Due to a limited number of words of length 17, 311 instead of 320 words were added

[8]`http://norvig.com/ngrams/count_1edit.txt`

[9]`http://www.cs.jhu.edu/~mdredze/data/`

[10]`https://healthlanguageprocessing.org/smm4h/challenge/`

---

[3]`www.reddit.com/r/cancer`

[4]`https://github.com/pushshift/api`

[5]Corpora and code are available on github `https://github.com/AnneDirkson`

Data set	Task	Size	Positive (%)	Negative (%)
Task 1 SMM4H 2019*	Presence adverse drug reaction	16,141	8.7	91.3
Task 4 SMM4H 2019* Flu vaccine	Personal health mention of flu vaccination	6,738	28.3	71.7
Flu Vaccination Tweets (Huang et al., 2017)	Relevance to topic flu vaccination	3,798	26.4	73.6
Twitter Health (Paul and Dredze, 2009)	Relevance to health	2,598	40.1	59.9
Task4 SMM4H 2019* Flu infection	Personal health mention of having flu	1,034	54.4	45.6
Zika Conspiracy Tweets (Dredze et al., 2016)	Contains pseudo-scientific information	588	25.9	74.1

Table 2: Six classification data sets of health-related Twitter data. *SMM4H: Social Media Mining 4 Health workshop

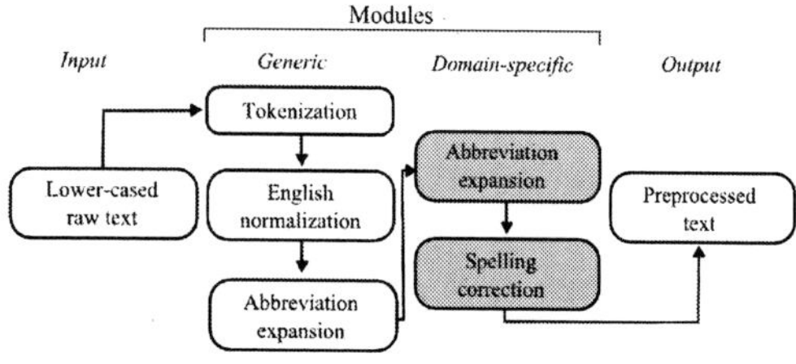

Figure 1: Sequential processing pipeline

4   Methods

Preprocessing   To protect the privacy of users, in-text person names were replaced as much as possible using a combination of the NLTK names corpus and part-of-speech tags (NNP and NNPS). Additionally, URLs and email addresses were replaced by the strings -url- and -email- using regular expressions. Furthermore, text was lower-cased and tokenized using NLTK. The first modules of the normalization pipeline of Sarker (2017) were employed: converting British to American English and normalizing generic abbreviations (see Figure 1). Some forum-specific additions were made: Gleevec (British variant: Glivec) was included in the British-American spelling conversion and one generic abbreviation expansion that clashed with a domain-specific one was substituted (i.e. 'temp' defined as *temperature* instead of *temporary*). Moreover, the abbreviations dictionary by Sarker (2017) was lower-cased. Lastly, domain-specific abbreviations were expanded with a lexicon of 42 non-ambiguous abbreviations, generated based on 500 randomly selected posts from the GIST forum and annotated by a domain expert and the first author. [11]

---

[11]This lexicon is shared on github `https://github.com/AnneDirkson`

Spelling correction   We used the method by Sarker (2017) as a baseline for spelling correction. Their method combines normalized absolute Levenshtein distance with Metaphone phonetic similarity and language model similarity. For the latter, distributed word representations (skip-gram word2vec) of three large Twitter data sets were used. In this paper, we used only the DIEGO LAB Drug Chatter Corpus (Sarker and Gonzalez, 2017a), as it was the only health-related corpus of the three. We also use a purely data-driven spelling correction method for comparison: Text-Induced Spelling Correction (TISC) developed by Reynaert (2005). It compares the anagrams of a token to those in a large corpus of text to correct mistakes. These two methods are compared with simple absolute and relative Levenshtein distance and weighted versions of both. To evaluate the spelling correction methods, the accuracy (i.e. the percentage of correct corrections) was used. The weights of the edits for weighted Levenshtein distance were computed using the log of the frequencies of the Norvig corpus. We used the log to ensure that a 10x more frequent error does not become 10x as cheap, as this would make infrequent errors too improbable. In order to make the weights inversely proportional to the frequencies and scale the weights between 0 and 1 with lower weights signifying lower costs for an edit, the following transformation of the log frequencies was used: Weight Edit Distance $= \frac{1}{1+log(frequency)}$.

Spelling mistake detection   We manually constructed a decision process, inspired by the work by Beeksma et al. (2019), for detecting spelling mistakes (See Figure 2). The decision process uses the corpus frequency relative to that of the token and the similarity to the token. The underlying idea is that if a word is either common within the domain-specific language or there is no similar enough candidate available, it is unlikely to be a mistake. A relative threshold enables us to capture more common mistakes.

To ensure generalisability, we opted for an unsupervised, data-driven method that does not rely on the construction of a specialized vocabulary. Candidates are considered in order of frequency. Of the candidates with the highest similarity score, the first is selected. The spelling correction ignores numbers and punctuation.

To optimize the decision process, a 10-fold cross validation grid search was conducted with a grid of 2 to 10 (steps of 1) for the minimum multiplication factor of the corpus frequency and a grid of 0.05 to 0.15 (steps of 0.01) for the minimum similarity. The choice of grid was based on previous work by Walasek (2016) and Beeksma et al. (2019). The loss function used to tune the parameters was the $F_{0.5}$ score, which places more weight on precision than the F_1 score. We believe it is more important to not alter correct terms, than to retrieve incorrect ones.

Spelling correction candidates For evaluating the mistake detection process, spelling correction candidates are derived from the data itself using the corpus frequency and similarity thresholds. For internal and external validation, candidates are also derived from the data itself. However, for comparing the spelling correction methods, the words of the specialized vocabulary for cancer forums (see section 3) were used as correction candidates in order to evaluate the methods independently of the vocabulary present in the data.

Internal validation The percentage of out-of-vocabulary (OOV) terms is used as an estimation of the quality of the data: less OOV-terms and thus more in-vocabulary (IV) terms is a proxy for cleaner data. As the correction candidates are derived from the data itself, one must note that words that are not part of CELEX may also be transformed from IV to OOV. The forum text was lemmatised prior to spelling correction. OOV analysis was done manually.

External validation Text classification was performed with default sklearn classifiers: Stochastic Gradient Descent (SGD), Multinomial Naive Bayes (MNB) and Linear Support Vector Machine (SVC). Uni-grams were used as features. A 10-fold cross-validation was used to determine the average score and paired t-test was applied to deter-

	Accuracy
Sarker's method	20.8 %
TISC	24.5 %
Absolute Edit distance (AE)	56.6 %
Relative Edit distance (RE)	56.6 %
Absolute Weighted Edit distance (AWE)	54.7 %
Relative Weighted Edit distance (RWE)	**62.3%**
Upper bound	84.9%

Table 3: Accuracy of spelling correction methods

mine significance of the absolute difference. Only the best performing classifier is reported per data set. For the shared tasks of the SMM4H workshop, only the training data was used.

To evaluate our method on generic social media text, we used the test set of the ACL W-NUT 2015 task (Baldwin et al., 2015). The test set consists of 1967 tweets with 2024 one-to-one, 704 one-to-many, and 10 many-to-one mappings. We did not need to use the training data, as our method is unsupervised. For comparison, the F_1 score on the W-NUT training data was 0.562.

5   Results

5.1   Spelling correction

The state-of-the-art method for generic social media performed poorly on medical social media with an accuracy of only 20.8% (see Table 3). A second established data-driven approach, TISC, also performed poorly (24.5%). The best performing baseline method on our spelling corpus was Relative Weighted Edit distance (RWE) (62.3%). As eight corrections did not occur in the CELEX, the upper bound was 84.9%.

One of the reasons for the low accuracy of Sarker's method may be the absence of correct terms (e.g. gleevec) in the language model it employs. This potential complication was already highlighted by Sarker (2017) in their own paper. Similarly, the large corpus of English news texts, which TISC relies on, may not contain the right terms or may not be comparable enough as a language model to our domain-specific data set.

In contrast, the key to the success of weighted edit distance methods is likely the incorporation of probabilities for 1-edit errors. This matches the intuition that certain errors are easier to make than others. For example, someone is more likely to wrongly spell sutent as sutant than as mutant (see Table 4). Such weighted methods indirectly integrate different types of possible errors, such as typo- and orthographical errors. The relative

Mistake	gllevec	stomack	sutant
Correct	gleevec*	stomach	sutent*
Sarker's method	clever	smack	mutant
TISC	gllevec	smack	dunant
AE	**gleevec**	**stomach**	mutant
RE	**gleevec**	**stomach**	mutant
AWE	**gleevec**	smack	**sutent**
RWE	**gleevec**	**stomach**	**sutent**

Table 4: Corrections made by spelling methods. *Gleevec and Sutent are important cancer medications for GIST patients

	$F_{0.5}$	F_1	Recall	Precision
CELEX	0.519	0.634	**1.0**	0.464
Decision process	**0.888**	**0.871**	0.844	**0.900**

Table 5: Results for mistake detection methods on the test set

False positives	oncologists	recruiter	angiogram
False negatives	norvay	stomach	vac

Table 6: Examples of errors of the decision process

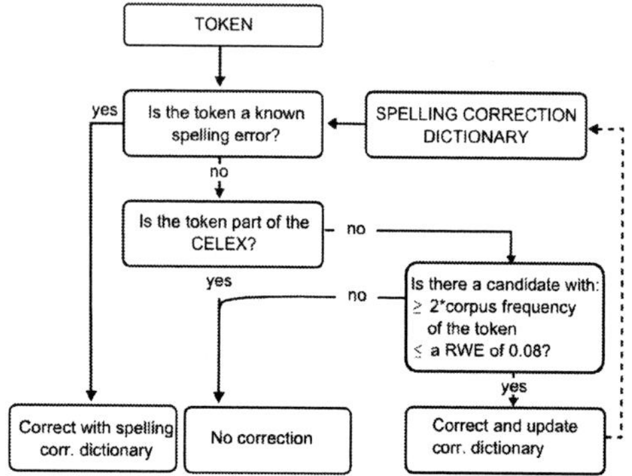

Figure 2: Decision process for spelling corrections. RWE: Relative Weighted Edit Distance

variant, as opposed to the absolute weighted edit distance, can counterbalance cheap deletions and additions, as can be seen for the mistake *stomack* (See Table 4).

5.2 Detecting spelling mistakes

The grid search results in two criteria for correction candidates: (1) a minimum of 2 times the relative corpus frequency of the token and (2) a maximum similarity score of 0.08 (see Figure 2). This combination attains the maximum $F_{0.5}$ score for all 10 folds.

On the test set, the decision process has an $F_{0.5}$ of 0.888. Its precision is high (0.90). Although the recall of a generic dictionary (i.e. CELEX) is maximal (1.0), its precision is low (0.464). This indicates, as hypothesized, that a dictionary-based method can retrieve more of the mistakes, but also will identify many correct terms as mistakes. Some examples of false positives were: 'oncologist', 'gleevec' and 'colonoscopy'. See Table 6 for some examples of errors made by our decision process.

The accuracy of the RWE method is further increased by 1.8% point by filtering the correction candidates using the preceding decision process, as is done in the full spelling module. The upper limit for spelling correction also increased from 84.9% to 92.5% by using candidates from the data instead of a specialized dictionary.

5.3 Effect on OOV rate

The reduction in OOV-terms was higher for the GIST (0.50%) than for the Reddit forum (0.27%) (See Figure 3). As expected, it appears that in-vocabulary terms are occasionally replaced with out-of-vocabulary terms, as the percentage of altered words is higher than the reduction in OOV (0.67% vs 0.50% for the GIST and 0.44% vs 0.27% for the Reddit forum).

Interestingly, the initial OOV count before spelling correction of the GIST forum is almost double that of the sub-reddit on cancer. This could be explained by the more specific nature of the forum: it may contain more words that are excluded from the dictionary, despite it being tailored to the cancer domain. This again underscores the limitations of dictionary-based methods.

Some of the most frequent corrections made in the GIST forum data were medical terms (e.g. gleevec, scan). Thus, although the overall reduction in OOV-terms may seem minor, our approach appears to target medical concepts, which are highly relevant for knowledge extraction tasks. Besides correcting mistakes in medical terms, our method also normalizes variants of medical terms (e.g. metastatic to metastasis). This is possibly a result of the corpus frequency comparison between tokens and candidates, which favors more prevalent variants.

Concerning the 50 most frequent remaining OOV terms, only a small proportion of them are in fact non-word spelling errors (e.g. 'wa'), although slang words (e.g 'ya') could arguably also be part of this category (see Table 7). A significant portion consists of real words (e.g. 'online', 'website', 'stressful') not present in the specialized dictio-

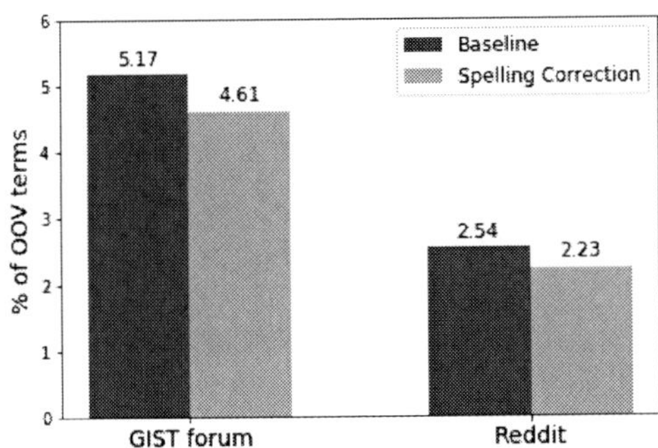

Figure 3: Percentage of OOV-terms in two cancer forums pre- and post-spelling correction.

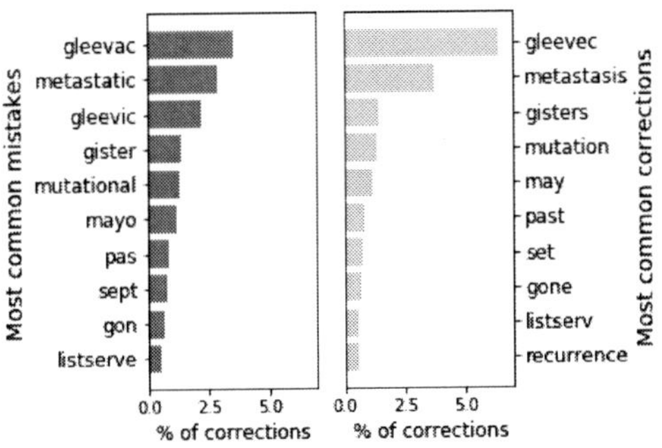

Figure 4: Most frequent mistakes and corrections in the GIST forum

nary. Upon manual inspection, the abbreviations frequently refer to treatments (e.g. 'rai'), mutation types (e.g. 'nf') or hospitals (e.g 'ucla'). Importantly, also some drug names are considered OOV (e.g. 'ativan'). Since they can be essential for downstream tasks, it is promising that they have not been altered by our method.

5.4 External evaluation

As can been seen in Table 8, the spelling correction does not lead to significant changes in the F_1 score for five of the six tasks. For the Twit-

	GIST forum	Reddit
Spelling error	3	1
Real word	11	21
Abbreviation	14	9
Slang	6	13
Name of person or hospital	14	2
Drug name	1	4
Not English	1	0
TOTAL	50	50

Table 7: Analysis of 50 most frequent remaining OOV in two cancer forums

ter Health classification task, the improvement is significant with a p-value of 0.041 according to a paired t-test.

In general, these changes are of the same order of magnitude as those made by the normalization pipeline of Sarker (2017). Moreover, the % of alterations due to spelling correction is comparable to that of the two cancer-related forums (see Figure 3). Although the overall classification accuracy on Task 1 of the SMM4H workshop is low, this is in line with the low F_1 score (0.522) of the best performing system on the comparable task in 2018 (Weissenbacher et al., 2018).

Neither the goal of the task, the relative amount of corrections nor the initial result seem to correlate with the change in F_1 score. Unlike in Sarker (2017), the improvements also do not seem to increase with the size of the data. The imbalance of the data may be associated with the change in accuracy to some extent: the two most balanced data sets show the largest increase (see Table 2). Further experiments would be necessary to elucidate if this is truly the case.

As can be seen in Table 9, our method does not perform well on generic social media text. In comparison, Sarker (2017)'s method attained state-of-the-art results with a F_1 of 0.836 on the ACL W-NUT 2015, but functioned poorly for medical social media (see Table 3). Thus, the success on one does not imply success on the other and consequently, normalisation of generic social media text and of domain-specific social media text appear different to the extent that they necessitate different approaches.

6 Discussion

Relative weighted edit distance outperforms both Sarker's method and other edit distance metrics with an accuracy of 62.3%. The accuracy is increased by a further 1.8% point if correction candidates are filtered with the criteria of the preceding decision process. This decision process is also capable of identifying mistakes with an $F_{0.5}$ of 0.888 and a high precision (0.90).

The spelling correction method led to an overall reduction in OOV-terms of 0.50% and 0.27% for two cancer-related forums. Although the reduction of OOV-terms may seem minor, relevant medical terms appear to be targeted (see Figure 4) and, additionally, many of the remaining OOV-terms are not spelling errors (see Table 7). Further-

Data set	Classifier	Prenorm F_1	Postnorm F_1	Postspell F_1	Change[+]	% of words corrected
Task 1 SMM4H 2019	SVC	0.410	0.413	0.417	+0.006	1.1
Task 4 SMM4H 2019 Flu vaccine	MNB	0.780	0.781	0.782	+0.001	0.47
Flu Vaccination Tweets	SVC	0.939	0.938	0.941	+0.002	0.83
Twitter Health	MNB	0.702	0.708	0.713	+0.010*	0.64
Task4 SMM4H 2019 Flu infection	MNB	0.784	0.792	0.795	+0.011	0.29
Zika Conspiracy Tweets	MNB	0.822	0.818	0.811	-0.011	1.1

Table 8: Mean classification accuracy before normalization (prenorm), after normalization (postnorm) and after spelling correction (postspell) for six health-related classification tasks. Only the results for the best performing classifier per data set are reported. MNB: Multinomial Naive Bayes; SVC: Linear Support Vector Classification. [+]Absolute change compared to prenorm.

	F_1	Precision	Recall
Sarker's method (Sarker, 2017)	**0.836**	**0.880**	0.796
IHS_RD (Supranovich and Patsepnia, 2015)	0.827	0.847	**0.808**
USZEGED (Berend and Tasnádi, 2015)	0.805	0.861	0.756
BEKLI (Beckley, 2015)	0.757	0.774	0.742
LYSGROUP (Doval Mosquera et al., 2015)	0.531	0.459	0.630
Our method	0.522	0.646	0.577

Table 9: Results for unconstrained systems of ACL W-NUT 2015

more, our method was designed to be conservative and to focus on precision to mitigate one of the major challenges of correcting errors in domain-specific data: the loss of information due to the 'correction' of correct domain-specific terms. The marginal change in task-based classification accuracy may be due to the fact that classification tasks do not rely strongly on individual terms, but on all words combined. This could also explain the lack of a correlation between the amount of alterations and the change in F_1 score. We plan to evaluate these results further by analysing both the corrections and the classification errors.

We speculate that our method will have a larger impact on named entity recognition (NER) tasks. Unfortunately, NER benchmarks for health-related social media are limited. We have investigated three relevant NER tasks that were publicly available: CADEC (Karimi et al., 2015), ADR-Miner (Nikfarjam et al., 2015), and the ADR extraction task of the SMM4H 2019. For all three tasks, extracted concepts could be matched exactly to the forum posts, thus negating the potential benefit of normalization. The exact matching can perhaps be explained by the fact that data collection and extraction from noisy text sources such as social media typically rely on keyword-based searching (Sarker and Gonzalez, 2017b).

Our study has a number of limitations. Firstly, the use of OOV-terms as a proxy for quality of the data relies heavily on the vocabulary that is chosen and, moreover, does not allow for differentiation between correct and incorrect substitutions. Consequently, we also test whether our method can improve classification accuracy on various tasks. Secondly, our method is currently targeted specifically at correcting non-word errors and is thus is unable to correct real word errors. Thirdly, our evaluation data set for developing our method is small: a larger evaluation data set would allow for more rigorous testing. Nonetheless, as far as we are aware, our corpora are the first for evaluating mistake detection and correction in a medical patient forum. We welcome comparable data sets sourced from various patient communities for further refinement and testing of our method.

7 Conclusion and future work

Our data-driven, unsupervised spelling correction can improve the quality of text data from medical forum posts from two cancer-related forums. Our method may also be useful for user-generated content in other highly specific and noisy domains, which contain many OOV compared to available dictionaries. Future work will include extending the pipeline with modules for named entity recognition, automated relation annotation and concept normalization.

8 Acknowledgements

We thank the SIDN fonds for financing this project and Abeed Sarker for his valuable feedback.

References

Timothy Baldwin, Marie-Catherine de Marneffe, Bo Han, Young-Bum Kim, Alan Ritter, and Wei Xu. 2015. Shared tasks of the 2015 workshop on noisy user-generated text: Twitter lexical normalization and named entity recognition. In *Proceedings of the Workshop on Noisy User-generated Text*, pages 126–135. Association for Computational Linguistics.

Russell Beckley. 2015. Bekli: A simple approach to twitter text normalization. In *Proceedings of the Workshop on Noisy User-generated Text*, pages 82–86, Beijing, China. Association for Computational Linguistics.

Merijn Beeksma, Suzan Verberne, Antal van den Bosch, Iris Hendrickx, Enny Das, and Stef Groenewoud. 2019. Predicting life expectancy with a recurrent neural network. *BMC Medical Informatics and Decision Making*. To appear.

Gábor Berend and Ervin Tasnádi. 2015. Uszeged: Correction type-sensitive normalization of english tweets using efficiently indexed n-gram statistics. In *Proceedings of the Workshop on Noisy User-generated Text*, pages 120–125, Beijing, China. Association for Computational Linguistics.

G. Burnage, R.H Baayen, R. Piepenbrock, and H. van Rijn. 1990. *CELEX: A Guide for Users*. Centre for Lexical Information.

Eleanor Clark and Kenji Araki. 2011. Text normalization in social media: Progress, problems and applications for a pre-processing system of casual english. *Procedia - Social and Behavioral Sciences*, 27:2 – 11. Computational Linguistics and Related Fields.

Yerai Doval Mosquera, Jesús Vilares, and Carlos Gómez-Rodríguez. 2015. Lysgroup: Adapting a spanish microtext normalization system to english. In *Proceedings of the Workshop on Noisy User-generated Text*, pages 99–105, Beijing, China. Association for Computational Linguistics.

Mark Dredze, David A Broniatowski, and Karen M Hilyard. 2016. Zika vaccine misconceptions: A social media analysis. *Vaccine*, 34(30):3441–2.

G Gonzalez-Hernandez, A Sarker, K O 'Connor, and G Savova. 2017. Capturing the Patient's Perspective: a Review of Advances in Natural Language Processing of Health-Related Text. *Yearbook of medical informatics*, pages 214–217.

Bo Han, Paul Cook, and Timothy Baldwin. 2012. Automatically constructing a normalisation dictionary for microblogs. In *Proceedings of the 2012 Joint Conference on Empirical Methods in Natural Language Processing and Computational Natural Language Learning*, pages 421–432, Jeju Island, Korea. Association for Computational Linguistics.

Bo Han, Paul Cook, and Timothy Baldwin. 2013. Lexical normalization for social media text. *ACM Transactions on Intelligent Systems and Technology*, 4(1):1–27.

Xiaolei Huang, Michael C. Smith, Michael Paul, Dmytro Ryzhkov, Sandra Quinn, David Broniatowski, and Mark Dredze. 2017. Examining Patterns of Influenza Vaccination in Social Media. In *AAAI Joint Workshop on Health Intelligence (W3PHIAI)*.

Ning Jin. 2015. NCSU-SAS-ning: Candidate generation and feature engineering for supervised lexical normalization. In *Proceedings of the Workshop on Noisy User-generated Text*, pages 87–92, Beijing, China. Association for Computational Linguistics.

Sarvnaz Karimi, Alejandro Metke-Jimenez, Madonna Kemp, and Chen Wang. 2015. Cadec: A corpus of adverse drug event annotations. *Journal of Biomedical Informatics*, 55:73–81.

Karen Kukich. 1992. Techniques for automatically correcting words in text. *ACM Comput. Surv.*, 24:377–439.

Kenneth H Lai, Maxim Topaz, Foster R Goss, and Li Zhou. 2015. Automated misspelling detection and correction in clinical free-text records. *Journal of Biomedical Informatics*, 55:188–195.

Samuel Leeman-Munk, James Lester, and James Cox. 2015. NCSU_SAS_SAM: Deep encoding and reconstruction for normalization of noisy text. In *Proceedings of the Workshop on Noisy User-generated Text*, pages 154–161, Beijing, China. Association for Computational Linguistics.

Wookhee Min and Bradford Mott. 2015. NCSU_SAS_WOOKHEE: A deep contextual long-short term memory model for text normalization. In *Proceedings of the Workshop on Noisy User-generated Text*, pages 111–119, Beijing, China. Association for Computational Linguistics.

Danielle L. Mowery, Brett R. South, Lee Christensen, Jianwei Leng, Laura Maria Peltonen, Sanna Salanterä, Hanna Suominen, David Martinez, Sumithra Velupillai, Noémie Elhadad, Guergana Savova, Sameer Pradhan, and Wendy W. Chapman. 2016. Normalizing acronyms and abbreviations to aid patient understanding of clinical texts: ShARe/CLEF eHealth Challenge 2013, Task 2. *Journal of Biomedical Semantics*.

National Cancer Institute. NCI Dictionary of Cancer Terms.

National Library of Medicine (US). RxNorm.

Azadeh Nikfarjam, Abeed Sarker, Karen O'Connor, Rachel Ginn, and Graciela Gonzalez. 2015. Pharmacovigilance from social media: mining adverse drug reaction mentions using sequence labeling with word embedding cluster features. *Journal of the*

American Medical Informatics Association: JAMIA, 22(3):671–81.

Albert Park, Andrea L Hartzler, Jina Huh, David W Mcdonald, and Wanda Pratt. 2015. Automatically Detecting Failures in Natural Language Processing Tools for Online Community Text. *J Med Internet Res*, 17(212).

Jon Patrick, Mojtaba Sabbagh, Suvir Jain, and Haifeng Zheng. 2010. Spelling correction in clinical notes with emphasis on first suggestion accuracy. In *2nd Workshop on Building and Evaluating Resources for Biomedical Text Mining*, pages 2–8.

Michael J Paul and Mark Dredze. 2009. A Model for Mining Public Health Topics from Twitter. Technical report, Johns Hopkins University.

Martin Reynaert. 2005. *Text-Induced Spelling Correction*. Ph.D. thesis, Tilburg University.

Abeed Sarker. 2017. A customizable pipeline for social media text normalization. *Social Network Analysis and Mining*, 7(1):45.

Abeed Sarker, Rachel Ginn, Azadeh Nikfarjam, Karen O'Connor, Karen Smith, Swetha Jayaraman, Tejaswi Upadhaya, and Graciela Gonzalez. 2015. Utilizing social media data for pharmacovigilance: A review. *Journal of Biomedical Informatics*, 54:202–212.

Abeed Sarker and Graciela Gonzalez. 2017a. A corpus for mining drug-related knowledge from Twitter chatter: Language models and their utilities. *Data in Brief*, 10:122–131.

Abeed Sarker and Graciela Gonzalez. 2017b. Hlp@upenn at semeval-2017 task 4a: A simple, self-optimizing text classification system combining dense and sparse vectors. In *Proceedings of the 11th International Workshop on Semantic Evaluation (SemEval-2017)*, pages 640–643, Vancouver, Canada. Association for Computational Linguistics.

Abeed Sarker, Karen O'Connor, Rachel Ginn, Matthew Scotch, Karen Smith, Dan Malone, and Graciela Gonzalez. 2016. Social Media Mining for Toxicovigilance: Automatic Monitoring of Prescription Medication Abuse from Twitter. *Drug Safety*, 39(3):231–240.

Dmitry Supranovich and Viachaslau Patsepnia. 2015. Ihs_rd: Lexical normalization for english tweets. In *Proceedings of the Workshop on Noisy User-generated Text*, pages 78–81, Beijing, China. Association for Computational Linguistics.

Suzan Verberne. 2002. Context-sensitive spell checking based on word trigram probabilities. Master's thesis, Radboud University.

Nicole Walasek. 2016. Medical Entity Extraction on Dutch forum data in the absence of labeled training data. Master's thesis, Radboud University.

Davy Weissenbacher, Abeed Sarker, Michael J. Paul, and Graciela Gonzalez-Hernandez. 2018. Overview of the third social media mining for health (smm4h) shared tasks at emnlp 2018. In *Proceedings of the 2018 EMNLP Workshop SMM4H: The 3rd Social Media Mining for Health Applications Workshop and Shared Task*, pages 13–16, Brussels, Belgium. Association for Computational Linguistics.

Q Zeng and T Tse. 2006. Exploring and developing consuming health vocabulary. *J Am Med Inform Assoc*, 13(1):24–29.

Xiaofang Zhou, An Zheng, Jiaheng Yin, Rudan Chen, Xianyang Zhao, Wei Xu, Wenqing Cheng, Tian Xia, and Simon Lin. 2015. Context-Sensitive Spelling Correction of Consumer-Generated Content on Health Care. *JMIR Med Inform 2015*, 3(3):e27.

Overview of the Fourth Social Media Mining for Health (#SMM4H) Shared Task at ACL 2019

Davy Weissenbacher[†], Abeed Sarker[†], Arjun Magge◇, Ashlynn Daughton[‡],
Karen O'Connor[†], Michael Paul[‡], Graciela Gonzalez-Hernandez[†]
[†]DBEI, Perelman School of Medicine, University of Pennsylvania, PA, USA
◇Biodesign Center for Environmental Health Engineering, Biodesign Institute,
Arizona State University, AZ, USA
[‡]Information Science University of Colorado Boulder, CO, USA
{dweissen, abeed, gragon}@pennmedicine.upenn.edu, amaggera@asu.edu
{mpaul, ashlynn.daughton}@colorado.edu

Abstract

The number of users of social media continues to grow, with nearly half of adults worldwide and two-thirds of all American adults using social networking on a regular basis[1]. Advances in automated data processing and NLP present the possibility of utilizing this massive data source for biomedical and public health applications, if researchers address the methodological challenges unique to this media. We present the Social Media Mining for Health Shared Tasks collocated with the ACL at Florence in 2019, which address these challenges for health monitoring and surveillance, utilizing state of the art techniques for processing noisy, real-world, and substantially creative language expressions from social media users. For the fourth execution of this challenge, we proposed four different tasks. Task 1 asked participants to distinguish tweets reporting an adverse drug reaction (ADR) from those that do not. Task 2, a follow-up to Task 1, asked participants to identify the span of text in tweets reporting ADRs. Task 3 is an end-to-end task where the goal was to first detect tweets mentioning an ADR and then map the extracted colloquial mentions of ADRs in the tweets to their corresponding standard concept IDs in the MedDRA vocabulary. Finally, Task 4 asked participants to classify whether a tweet contains a personal mention of one's health, a more general discussion of the health issue, or is an unrelated mention. A total of 34 teams from around the world registered and 19 teams from 12 countries submitted a system run. We summarize here the corpora for this challenge which are freely available at https://competitions.codalab.org/competitions/22521, and present an overview of the methods and the results of the competing systems.

---

[1]Pew Research Center. Social Media Fact Sheet. 2017. [Online]. Available: http://www.pewinternet.org/fact-sheet/social-media/

1 Introduction

The intent of the #SMM4H shared tasks series is to challenge the community with Natural Language Processing tasks for mining relevant data for health monitoring and surveillance in social media. Such challenges require processing imbalanced, noisy, real-world, and substantially creative language expressions from social media. The competing systems should be able to deal with many linguistic variations and semantic complexities in the various ways people express medication-related concepts and outcomes. It has been shown in past research (Liu et al., 2011; Giuseppe et al., 2017) that automated systems frequently under-perform when exposed to social media text because of the presence of novel/creative phrases, misspellings and frequent use of idiomatic, ambiguous and sarcastic expressions. The tasks act as a discovery and verification process of what approaches work best for social media data.

As in previous years, our tasks focused on mining health information from Twitter. This year we challenged the community with two different problems. The first problem focuses on performing pharmacovigilance from social media data. It is now well understood that social media data may contain reports of adverse drug reactions (ADRs) and these reports may complement traditional adverse event reporting systems, such as the FDA adverse event reporting system (FAERS). However, automatically curating reports from adverse reactions from Twitter requires the application of a series of NLP methods in an end-to-end pipeline (Sarker et al., 2015). The first three tasks of this year's challenge represent three key NLP problems in a social media based pharmacovigilance pipeline — (i) automatic classification of ADRs, (ii) extraction of spans of ADRs and (iii) normal-

21

ization of the extracted ADRs to standardized IDs.

The second problem explores the generalizability of predictive models. In health research using social media, it is often necessary for researchers to build individual classifiers to identify health mentions of a particular disease in a particular context. Classification models that can generalize to different health contexts would be greatly beneficial to researchers in these fields (e.g., (Payam and Eugene, 2018)), as this would allow researchers to more easily apply existing tools and resources to new problems. Motivated by these ideas, Task 4 was testing tweet classification methods across diverse health contexts, so the test data included a very different health context than the training data. This setting measures the ability of tweet classifiers to generalize across health contexts.

The fourth iteration of our series follows the same organization as previous iterations. We collected posts from Twitter, annotated the data for the four tasks proposed and released the posts to the registered teams. This year, we conducted the evaluation of all participating systems using Codalab, an open source platform facilitating data science competitions. The performances of the systems were compared on a blind evaluations sets for each task.

All teams registered were allowed to participate to one or multiple tasks. We provided the participants with two sets of data for each task, a training and a test set. Participants had a period of six weeks, from March 5th to April 15th, for training their systems on our training sets, and 4 days, from the 16th to 20th of April, for calibrating their systems on our test sets and submitting their predictions. In total 34 teams registered and 19 teams submitted at least one run (each team was allowed to submit, at most, three runs per task). In detail, we received 43 runs for task 1, 24 for task 2, 10 for task 3 and 15 for task 4. We briefly describe each task and their data in section 2, before discussing the results obtained in section 3.

2 Task Descriptions

2.1 Tasks

Task 1: Automatic classification of tweets mentioning an ADR. This is a binary classification task for which systems are required to predict if a tweet mentions an ADR or not. In an end-to-end social media based pharmacovigilance pipeline, such a system is needed after data collection to filter out the large volume of medication-related chatter that is not a mention of an ADR. This task is a rerun of the popular classification task organized in past years.

Task 2: Automatic extraction of ADR mentions from tweets. This is a named entity recognition (NER) task that typically follows the ADR classification step (Task 1) in an ADR extraction pipeline. Given a set of tweets containing drug mentions and potentially containing ADRs, the objective was to determine the span of the ADR mention, if any. ADRs are rare events making ADR classification a challenging task with an F1-score in the vicinity of 0.5 (based on previous shared task results (Weissenbacher et al., 2018)) for the ADR class. The dataset for the ADR extraction task contains tweets that are both positive and negative for the presence of ADRs. This allowed participants to choose to train their systems on either the set of tweets containing ADRs or include tweets that were negative for the presence of ADRs.

Task 3: Automatic extraction of ADR mentions and normalization of extracted ADRs to MedDRA preferred term identifiers. This is an extension of Task 2 consisting of the combination of NER and entity normalization tasks: a named entity resolution task. In this task, given the same set of tweets as in Task 2, the objective was to extract the span of an ADR mention and to normalize it to MedDRA identifiers [2]. MedDRA (Medical Dictionary for Regulatory Activities), which is the standard nomenclature for monitoring medical products, and includes diseases, disorders, signs, symptoms, adverse events or adverse drug reactions. For the normalization task, MedDRA version 21.1 was used, containing 79,507 lower level terms (LLTs) and 23,389 respective preferred terms (PTs).

Task 4: Automatic classification of personal mentions of health. In this binary classification task, the systems were required to distinguish tweets of personal health status or opinions across different health domains. The proposed task was intended to provide a baseline understanding of the ability to identify personal health mentions in a generalized context.

---

[2]`https://www.meddra.org/`   Accessed: 05/13/2019.

2.2 Data

All corpora were composed of public tweets downloaded using the official streaming API provided by Twitter and made available to the participants in accordance with Twitter's data use policy. This study received an exempt determination by the Institutional Review Board of the University of Pennsylvania.

Task 1. For training, participants were provided with all the tweets from the #SMM4H 2017 shared tasks (Sarker et al., 2018), which are publicly available at: `https://data.mendeley.com/datasets/rxwfb3tysd/2`. A total of 25,678 tweets were made available for training. The test set consisted of 4575 tweets with 626 (13.7%) tweets representing ADRs. The evaluation metric for this task was micro-averaged F1-score for the ADR class.

Task 2. Participants of Task 2 were provided with a training set containing 2276 tweets which mentioned at least one drug name. The dataset contained 1300 tweets that were positive for the presence of ADRs and 976 tweets that were negative. Participants were allowed to include additional negative instances from Task 1 for training purposes. Positive tweets were annotated with the start and end indices of the ADRs and the corresponding span text in the tweets. The evaluation set contained 1573 tweets, 785 and 788 tweets were positive and negative for the presence of ADRs respectively. The participants were asked to submit outputs from their systems that contained the predicted start and end indices of ADRs. The participants' submissions were evaluated using standard strict and overlapping F1-scores for extracted ADRs. Under strict mode of evaluation, ADR spans were considered correct only if both start and end indices matched with the indices in our gold standard annotations. Under overlapping mode of evaluation, ADR spans were considered correct only if spans in predicted annotations overlapped with our gold standard annotations.

Task 3. Participants were provided with the same training and evaluation datasets as in Task 2. However, the datasets contained additional columns for the MedDRA annotated LLT and PT identifiers for each ADR mention. In total, of the 79,507 LLT and 23,389 PT identifiers available in MedDRA, the training set of 2276 tweets and 1832 annotated ADRs contained 490 unique LLT iden-

tifiers and 327 unique PT identifiers. The evaluation set contained 112 PT identifiers that were not present as part of the training set. The participants were asked to submit outputs containing the predicted start and end indices of ADRs and respective PT identifiers. Although the training dataset contained annotations at the LLT level, the performance was only evaluated at the higher PT level. The participants' submissions were evaluated using standard strict and overlapping F-scores for extracted ADRs and respective MedDRA identifiers. Under strict mode of evaluation, ADR spans were considered correct only if both start and end indices matched along with matching MedDRA PT identifiers. Under overlapping mode of evaluation, ADR spans were considered correct only if spans in predicted ADRs overlapped with gold standard ADR spans in addition to matching MedDRA PT identifiers.

Task 4 Data. Participants were provided training data from one disease domain, influenza, across two contexts, being sick and getting vaccinated, both annotated for personal mentions: the user is personally sick or the user has been personally vaccinated. Test data included new tweets of personal health mentions about influenza and tweets from an additional disease domain, Zika virus, with two different contexts, the user is changing their travel plans in response to Zika concerns, or the user is minimizing potential mosquito exposure due to Zika concerns.

2.3 Annotation and Inter-Annotator Agreements

Two annotators with biomedical education and both experienced in Social Media research tasks manually annotated the corpora for tasks 1, 2 and 3. Our annotators independently dual-annotated each test sets to insure the quality of our annotations. Disagreement were resolved after an adjudication phase between our two annotators. On task 1, the classification task, the inter annotator-agreement (IAA) was high with a Cohens Kappa = 0.82. On task 2, the information extraction task, IAAs were good with and an F1-score of 0.73 for strict agreement, and 0.85 for overlapping agreement[3]. On task 3, our annotators double annotated

---

[3] Since task 2 is a named-entity recognition task, we followed the recommendations of (Hripcsak and Rothschild, 2005) and used precision and recall metrics to estimate the inter-annotator rate.

535 of the extracted ADR terms and normalized them to MedDRA lower lever terms (LLT). They achieved an agreement accuracy of 82.6%. After converting the LLT to their corresponding preferred term (PT) in MedDRA, which is the coding the task was scored against, accuracy improved to 87.7%[4].

The annotation process followed for task 4 was slightly different due to the nature of the task. We obtained the two datasets of our training set, focusing on flu vaccination and flu infection, from (Huang et al., 2017) and (Lamb et al., 2013) respectively. Huang et al. (Huang et al., 2017) used mechanical turk to crowdsource labels (Fleiss' kappa = 0.793). Lamb et al. (Lamb et al., 2013) did not report their labeling procedure or annotator agreement metrics, but do report annotation guidelines[5]. A few of the tweets released by Lamb et al. appeared to be mislabeled and were corrected in accordance with the annotation guidelines defined by the authors. We obtained the test data for task 4 by compiling three datasets. For the dataset related to travel changes due to Zika concerns, we selected a subset of data already available from (Daughton and Paul, 2019). Initial labeling of these tweets was performed by two annotators with a public health background (Cohen's kappa = 0.66). We reuse the original annotations for this dataset without changes. For the mosquito exposure dataset, tweets were labeled by one annotator with public health knowledge and experienced with social media, and then verified by a second annotator with similar experience. The additional set of data on personal exposure to Influenza were obtained from a separate group, who used an independent labeling procedure.

3   Results

The challenge received a solid response with 19 teams from 12 countries (7 from North America, 1 from South America, 6 from Asia and 5 from Europe) submitting 92 runs in total in one or more tasks. We present an overview of all architectures competing in the different tasks in Table 1, 2, 3, 4. We also list in these tables the external resources competitors integrated for improving

the pre-training of their systems or for embedding high-level features to help decision-making.

The overview of all architectures is interesting in two ways. First, this challenge confirms the tendency of the community to abandon traditional Machine Learning systems based on hand-crafted features for deep learning architectures capable of discovering the features relevant for the task at hand from pre-trained embeddings. During the challenge, when participants implemented traditional systems, such as SVM or CRF, they used such systems as baselines and, observing significant differences of performances with systems based on deep learning on their validation sets, most of them did not submit their predictions as official runs. Second, while last year convolutional or recurrent neural networks "fed" with pre-trained word embeddings learned on local windows of words (e.g. word2vec, GloVe) were the most popular architectures, this year we can see a clear dominance of neural architectures using word embeddings pre-trained with the Bidirectional Encoder Representations from Transformers (BERT) proposed by (Devlin et al., 2018), or fine-tuning these words embeddings on our training corpora. BERT allows to compute words embeddings based on the full context of sentences and not only on local windows.

A notable result from task 1-3 is that, despite an improvement in performances for the detection of ADRs, their resolution remains challenging and will require further research. The participants largely adopted contextual word-embeddings during this challenge, a choice rewarded by new records in performances during the task 1, the only task reran from last years. The performances increased from .522 F1-score (.442 P, .636 R) (Weissenbacher et al., 2018) to .646 F1-score (0.608 P, 0.689 R) for the best systems of each years. However, with a strict matching F1-score of .432 (.362 P, .535 R) for the best system, the performances obtained in task 3 for ADRs resolution are still low and human inspection is still required to make use of the data extracted automatically. As shown by the best score of .887 Accuracy obtained on the ADR normalization in task 3 ran during #SMM4H in 2017 (Sarker et al., 2018)[6], once ADRs are extracted, the normalization of the ADRs can be per-

---

[4]We measured agreement using accuracy instead of Cohens Kappa because, with greater than 70,000 LLTs for the annotators to choose from, agreement due to chance is expected to be small.

[5]We used the awareness vs. infection labels as defined in (Lamb et al., 2013).

[6]Organizers of the task 3 ran during #SMM4H 2017 provided participants with manually curated expressions referring to ADRs and participants had to map them to their corresponding preferred terms in MeDRA.

formed with a good reliability. However errors are made during all steps of the resolution — detection, extraction, normalization — and their overall accumulation render current automatic systems inefficient. Note that bulk of the errors are made during the extraction of the ADRs, as shown by the low strict F1-score of the best system in task 2, .464 F1-score (.389P, .576 R).

For task 4, we were especially interested in the generalizability of first person health classifiers to a domain separate from that of the training data. We find that, on average, teams do reasonably well across the full test dataset (average F1-score: 0.70, range: 0.41-0.87). Unsurprisingly, classifiers tended to do better on a test set in the same domain as the training dataset (context 1, average F1-score: 0.82) and more modestly on the Zika travel and mosquito datasets (average F1-score: 0.40 and 0.52, respectively). Interestingly, in all contexts, precision was higher than recall. We note that both the training and the testing data were limited in quantity, and that classifiers would likely improve with more data. However, in general, it is encouraging that classifiers trained in one health domain can be applied to separate health domains.

4   Conclusion

In this paper we presented an overview of the results of #SMM4H 2019 which focuses on a) the resolution of adverse drug reaction (ADR) mentioned in Twitter and b) the distinction between tweets reporting personal health status form opinions across different health domains. With a total of 92 runs submitted by 19 teams, the challenge was well attended. The participants, in large part, opted for neural architectures and integrated pre-trained word-embedding sensitive to their contexts based on the recent Bidirectional Encoder Representations from Transformers. Such architectures were the most efficient on our four tasks. Results on tasks 1-3 show that, despite a continuous improvement of performances in the detection of tweets mentioning ADRs over the past years, their end-to-end resolution still remain a major challenge for the community and an opportunity for further research. Results of task 4 were more encouraging, with systems able to generalized their predictions over domains not present in their training data.

References

Ashlynn R. Daughton and Michael J. Paul. 2019. Identifying protective health behaviors on twitter: Observational study of travel advisories and zika virus. *Journal of Medical Internet Research. In Press.*

Jacob Devlin, Ming-Wei Chang, Kenton Lee, and Kristina Toutanova. 2018. BERT: pre-training of deep bidirectional transformers for language understanding. *CoRR*, abs/1810.04805.

Rizzo Giuseppe, Pereira Bianca, Varga Andrea, van Erp Marieke, and Elizabeth Cano Basave Amparo. 2017. Lessons learnt from the named entity recognition and linking (neel) challenge series. *Semantic Web Journal*, 8(5):667–700.

George Hripcsak and Adam S Rothschild. 2005. Agreement, the f-measure, and reliability in information retrieval. *Journal of the American Medical Informatics Association*, 12(3):296–298.

Xiaolei Huang, Michael C. Smith, Michael J. Paul, Dmytro Ryzhkov, Sandra C. Quinn, David A. Broniatowski, and Mark Dredze. 2017. Examining patterns of influenza vaccination in social media. In *AAAI Joint Workshop on Health Intelligence (W3PHIAI)*.

Alex Lamb, Michael J. Paul, and Mark Dredze. 2013. Separating fact from fear: Tracking flu infections on twitter. In *Proceedings of Conference of the North American Chapter of the Association for Computational Linguistics: Human Language Technologies (NAACL-HLT 2013)*.

Xiaohua Liu, Shaodian Zhang, Furu Wei, and Ming Zhou. 2011. Recognizing named entities in tweets. In *Proceedings of the 49th Annual Meeting of the Association for Computational Linguistics: Human Language Technologies - Volume 1*, HLT '11, pages 359–367. Association for Computational Linguistics.

Zulfat Miftahutdinov, Elena Tutubalina, and Alexander Tropsha. 2017. Identifying disease-related expressions in reviews using conditional random fields. In *Proceedings of International Conference on Computational Linguistics and Intellectual Technologies Dialog*, volume 1, pages 155–167.

Azadeh Nikfarjam, Abeed Sarker, Karen O'connor, Rachel Ginn, and Graciela Gonzalez-Hernandez. 2015. Pharmacovigilance from social media: mining adverse drug reaction mention using sequence labeling with word embedding cluster features. *Journal of the American Medical Informatics Association*, 22(3):671–681.

Karisani Payam and Agichtein Eugene. 2018. Did you really just have a heart attack? towards robust detection of personal health mentions in social media. In *Proceedings of the 2018 World Wide Web Conference on World Wide Web*, pages 137–146.

Abeed Sarker, Maksim Belousov, Jasper Friedrichs, Kai Hakala, Svetlana Kiritchenko, Farrokh Mehryary, Sifei Han, Tung Tran, Anthony Rios, Ramakanth Kavuluru, Berry de Bruijn, Filip Ginter, Debanjan Mahata, Saif Mohammad, Goran Nenadic, and Graciela Gonzalez-Hernandez. 2018. Data and systems for medication-related text classification and concept normalization from twitter: insights from the social media mining for health (smm4h)-2017 shared task. *J Am Med Inform Assoc*, 25(10):1274–1283.

Abeed Sarker, Rachel Ginn, Azadeh Nikfarjam, Karen OConnor, Karen Smith, Swetha Jayaraman, Tejaswi Upadhaya, and Graciela Gonzalez. 2015. Utilizing social media data for pharmacovigilance: A review. *Journal of Biomedical Informatics*, 54:202 – 212.

Abeed Sarker and Graciela Gonzalez. 2017. A corpus for mining drug-related knowledge from twitter chatter: Language models and their utilities. *Data in Brief*, 10:122–131.

Abeed Sarker and Graciela Gonzalez-Hernandez. 2015. Portable automatic text classification for adverse drug reaction detection via multi-corpus training. *Journal of biomedical informatics*, 53:196–207.

Davy Weissenbacher, Abeed Sarker, Michael J Paul, and Graciela Gonzalez-Hernandez. 2018. Overview of the third social media mining for health (smm4h) shared tasks at emnlp 2018. In *in Proceedings of the 2018 EMNLP Workshop SMM4H: The 3rd Social Media Mining for Health Applications Workshop and Shared Task*, pages 13–16.

Abbreviations

FF: Feedforward
CNN: Convolutional Neural Network
BiLSTM: Bidirectional Long Short-Term Memory
SVM: Support Vector Machine
CRF: Conditional Random Field
POS: Part-Of-Speech
RNN: Recurrent Neural Network

Rank	Team	System details
1	ICRC	*Architecture:* BERT + FF + Softmax
		Details: lexicon features (pairs of drug-ADR)
		Resources: SIDER
2	UZH	*Architecture:* ensemble of BERT & C_CNN + W_BiLSTM (+ CRF)
		Details: multi-task-learning
		Resources: CADEC corpus
3	MIDAS@IIITD	*Architecture:* 1. BERT 2. ULMFit 3. W_BiLSTM
		Details: BERT + GloVe + Flair
		Resources: additional corpus (Sarker and Gonzalez-Hernandez, 2015)
4	KFU NLP	*Architecture:* BERT + logistic regression
		Details: BioBERT
5	CLaC	*Architecture:* Bert + W_BiLSTM + attention + softmax + SVM
		Details: BERT, Word2Vec, Glove, embedded features
		Resources: POS, modality, ADR list
6	THU_NGN	*Architecture:* C_CNN + W_BiLSTM + features + Multi-Head attention + Softmax
		Details: Word2Vec, POS, ELMo
		Resources: sentiment Lexicon, SIDER, CADEC
7	BigODM	*Architecture:* ensemble of SVMs
		Resources: Word Embeddings
8	UMich-NLP4Health	*Architecture:* 1. W_BiLSTM + attention + softmax; 2. W_CNN + BiLSTM + softmax; 3. SVM
		Details: GloVe, POS, case
		Resources: Metamap, cTAKES, CIDER
9	TMRLeiden	*Architecture:* ULMfit
		Details: Flair + Glove + Bert; transfer learning
		Resources: external corpus (Sarker and Gonzalez, 2017)
10	CIC-NLP	*Architecture:* C_BiLSTM + W_FF + LSTM + FF
		Details: GloVe + BERT
12	SINAI	*Architecture:* 1. SVM 2. CNN + Softmax
		Details: GloVe
		Resources: MetaMap
13	nlp-uned	*Architecture:* W_BiLSTM + Sigmoid
		Details: GloVe
14	ASU BioNLP	*Architecture:* 1. Lexicon; 2. BioBert
		Details: Lexicon learned with Logistic regression model
15	Klick Health	*Architecture:* ELMo + FF + Softmax
		Details: Lexicons
		Resources: MedDRA, Consumer Health Vocabulary, (Nikfarjam et al., 2015)
16	GMU	*Architecture:* encoder-decoder (W_biLSTM + attention)
		Details: Glove
		Resources: #SMM4H 2017-2018, UMLS

Table 1: Task 1. System and resource descriptions for ADR mentions detection in tweets[7].

[8] We use C_BiLSMT and C_CNN to denote bidirectonal LSTMs or CNNs encoding sequences of characters, W_BiLSTM and W_FF to denote bidirectional LSTMs or Feed Forward encoders of word embeddings.

Rank	Team	System details
1	KFU NLP	*Architecture:* ensemble of BioBERT + CRF
		Details: BioBERT
		Resources: external dictionaries (Miftahutdinov et al., 2017);
		CADEC, PsyTAR, TwADR-L corpora; #SMM4H 2017
2	THU_NGN	*Architecture:* C_CNN + W_BiLSTM + features + Multi-Head self-attention + CRF
		Details: Word2Vec, POS, ELMo
		Resources: sentiment Lexicon, SIDER, CADEC
3	MIDAS@IIITD	*Architecture:* W_BiLSTM + CRF
		Details: BERT + GloVe + Flair
4	TMRLeiden	*Architecture:* BERT + Flair
		Details: Flair + Glove + Bert; transfer learning
5	ICRC	*Architecture:* BERT + CRF
		Resources: SIDER
6	GMU	*Architecture:* C_biLSTM + W_biLSTM + CRF
		Details: Glove
		Resources: #SMM4H 2017-2018, UMLS
7	HealthNLP	*Architecture:* W_BiLSTM + CRF
		Details: Word2vec, BERT, ELMo, POS
		Resources: external dictionaries
8	SINAI	*Architecture:* CRF
		Details: GloVe
		Resources: MetaMap
9		*Architecture:* BiLSTM + CRF
		Details: Word2Vec
		Resources: MIMIC-III
10	Klick Health	*Architecture:* Similarity
		Details: Lexicons
		Resources: MedDRA, Consumer Health Vocabulary, (Nikfarjam et al., 2015)

Table 2: Task 2. System and resource descriptions for ADR mentions extraction in tweets

Rank	Team	System details
1	KFU NLP	*Architecture:* BioBERT + softmax
2	myTomorrows-TUDelft	*Architecture:* ensemble RNN & Few-Shot Learning
		Details: Word2Vec
		Resources: MedDRA, Consumer Health Vocabulary, UMLS
3	TMRLeiden	*Architecture:* BERT + Flair + RNN
		Details: Flair + Glove + Bert; transfer learning
		Resources: Consumer Health Vocabulary
4	GMU	*Architecture:* encoder-decoder (W_biLSTM + attention)
		Details: Glove
		Resources: #SMM4H 2017-2018, UMLS

Table 3: Task 3. System and resource descriptions for ADR mentions resolution in tweets.

Rank	Team	System details
1	UZH	*Architecture:* ensemble BERT
		Resources: CADEC corpus
2	ASU1	*Architecture:* BioBERT + FF
		Resources: Word2vec, manually compiled list, ConceptNet
4	MIDAS@IIITD	*Architecture:* BERT; W_BiLSTM
		Details: BERT + GloVe + Flair
5	TMRLeiden	*Architecture:* ULMfit
		Details: Flair + Glove + Bert; transfer learning
		Resources: external corpus (Payam and Eugene, 2018)
6	CLaC	*Architecture:* Bert + W_BiLSTM + attention + softmax + SVM
		Details: BERT, Word2Vec, Glove, embedded features
		Resources: POS, modality, ADR list

Table 4: Task 4. System and resource descriptions for detection of personal mentions of health in tweets.

Team	F1	P	R
ICRC	**0.6457**	0.6079	0.6885
UZH	0.6048	0.6478	0.5671
MIDAS@IIITD	0.5988	0.6647	0.5447
KFU NLP	0.5738	**0.6914**	0.4904
CLaC	0.5738	0.5427	0.6086
THU_NGN	0.5718	0.4667	**0.738**
BigODM	0.5514	0.4762	0.655
UMich-NLP4Health	0.5369	0.5654	0.5112
TMRLeiden	0.5327	0.6419	0.4553
CIC-NLP	0.5209	0.6203	0.4489
UChicagoCompLx	0.4993	0.4574	0.5495
SINAI	0.4969	0.5517	0.4521
nlp-uned	0.4723	0.5244	0.4297
ASU BioNLP	0.4317	0.3223	0.6534
Klick Health	0.4099	0.5824	0.3163
GMU	0.3587	0.4526	0.2971

Table 5: System performances for each team for task 1 of the shared task. F1-score, Precision and Recall over the ADR class are shown. Top scores in each column are shown in bold.

Team	Relaxed			Strict		
	F1	P	R	F1	P	R
KFU NLP	**0.658**	0.554	**0.81**	**0.464**	0.389	**0.576**
THU_NGN	0.653	0.614	0.697	0.356	0.328	0.388
MIDAS@IIITD	0.641	0.537	0.793	0.328	0.274	0.409
TMRLeiden	0.625	0.555	0.715	0.431	0.381	0.495
ICRC	0.614	0.538	0.716	0.407	0.357	0.474
GMU	0.597	0.596	0.599	0.407	0.406	0.407
HealthNLP	0.574	**0.632**	0.527	0.336	0.37	0.307
SINAI	0.542	0.612	0.486	0.36	**0.408**	0.322
ASU BioNLP	0.535	0.415	0.753	0.269	0.206	0.39
Klick Health	0.396	0.416	0.378	0.194	0.206	0.184

Table 6: System performances for each team for task 2 of the shared task. (Strict/Relaxed) F1-score, Precision and Recall over the ADR mentions are shown. Top scores in each column are shown in bold.

Team	Relaxed			Strict		
	F1	P	R	F1	P	R
KFU NLP	**0.432**	0.362	**0.535**	**0.344**	0.288	**0.427**
myTomorrows-TUDelft	0.345	0.336	0.355	0.244	0.237	0.252
TMRLeiden	0.312	**0.37**	0.27	0.25	**0.296**	0.216
GMU	0.208	0.221	0.196	0.109	0.116	0.102

Table 7: System performances for each team for task 3 of the shared task. (Strict/Relaxed) F1-score, Precision and Recall over the ADR resolution are shown. Top scores in each column are shown in bold.

Team	Acc	F1	P	R
Health concerns in all contexts				
UZH	**0.8772**	**0.8727**	0.8392	**0.9091**
ASU1	0.8456	0.8036	**0.9783**	0.6818
UChicagoCompLx	0.8316	0.7913	0.9286	0.6894
MIDAS@IIITD	0.8211	0.783	0.8932	0.697
TMRLeiden	0.793	0.7256	0.9398	0.5909
CLaC	0.6386	0.4607	0.7458	0.3333
Health concerns in Context 1: Flu virus (infection/vaccination)				
UZH	0.9438	0.9474	0.9101	0.9878
UChicagoCompLx	0.925	0.9231	0.973	0.878
ASU1	0.925	0.9221	0.9861	0.8659
MIDAS@IIITD	0.8875	0.88	0.9706	0.8049
TMRLeiden	0.8625	0.8493	0.9688	0.7561
CLaC	0.6625	0.5645	0.8333	0.4268
Health concerns in Context 2: Zika virus, travel plans changes				
UZH	0.7536	0.7385	0.7059	0.7742
MIDAS@IIITD	0.6667	0.5818	0.6667	0.5161
ASU1	0.6957	0.5116	0.9167	0.3548
UChicagoCompLx	0.6377	0.4681	0.6875	0.3548
TMRLeiden	0.6377	0.4186	0.75	0.2903
CLaC	0.5362	0.2	0.4444	0.129
Health concerns in Context 3: Zika virus, reducing mosquito exposure				
UZH	0.8393	0.7692	0.75	0.7895
MIDAS@IIITD	0.8214	0.6667	0.9091	0.5263
ASU1	0.8036	0.5926	1.0	0.4211
UChicagoCompLx	0.8036	0.5926	1.0	0.4211
TMRLeiden	0.7857	0.5385	1.0	0.3684
CLaC	0.6964	0.3704	0.625	0.2632

Table 8: System performances for each team for task 4 of the shared task. Accuracy, F1-score, Precision and Recall over the personal mentions are shown. Top scores in each column are shown in bold.

MedNorm: A Corpus and Embeddings for Cross-terminology Medical Concept Normalisation

Maksim Belousov
School of Computer Science
The University of Manchester
United Kingdom

William G. Dixon
Centre for Epidemiology
Versus Arthritis
The University of Manchester
United Kingdom

Goran Nenadic
School of Computer Science
The University of Manchester
United Kingdom

{maksim.belousov, will.dixon, g.nenadic}@manchester.ac.uk

Abstract

The medical concept normalisation task aims to map textual descriptions to standard terminologies such as SNOMED-CT or MedDRA. Existing publicly available datasets annotated using different terminologies cannot be simply merged and utilised, and therefore become less valuable when developing machine learning-based concept normalisation systems. To address that, we designed a data harmonisation pipeline and engineered a corpus of 27,979 textual descriptions simultaneously mapped to both MedDRA and SNOMED-CT, sourced from five publicly available datasets across biomedical and social media domains. The pipeline can be used in the future to integrate new datasets into the corpus and also could be applied in relevant data curation tasks. We also described a method to merge different terminologies into a single concept graph preserving their relations and demonstrated that representation learning approach based on random walks on a graph can efficiently encode both hierarchical and equivalent relations and capture semantic similarities not only between concepts inside a given terminology but also between concepts from different terminologies. We believe that making a corpus and embeddings for cross-terminology medical concept normalisation available to the research community would contribute to a better understanding of the task.

1 Introduction

The medical concept normalisation task aims to assign a corresponding identifier from a standard terminology to text descriptions. Depending on the domain, descriptions may vary from formal medical jargon terms (e.g. *"Dizziness"*) to more informal and colloquial expressions that rather explain how the patient feels (e.g. *"everything that surrounds me is circling or rolling"*, *"kept bumping into things"*). There are multiple terminologies of medical concepts that are commonly used for mapping, such as SNOMED-CT (Systematized Nomenclature of Medicine - Clinical Terms) (Stearns et al., 2001) and MedDRA (Medical Dictionary for Regulatory Activities) (Brown et al., 1999). The Unified Medical Language System (UMLS) (Schuyler et al., 1993) integrates concepts from various biomedical vocabularies and lexicons, including SNOMED-CT and MedDRA. Each concept is represented by its Concept Unique Identifier (CUI). Clinicians choose the most suitable terminology based on their particular case or application. Hence, when creating corpora with annotated medical concepts, there is no general agreement on which terminology to use or which annotation guidelines to follow. Also, variety of available concepts in terminologies (e.g. over 70,000 lowest level terms in MedDRA and over 350,000 concepts in SNOMED-CT) makes it harder to achieve high agreement between annotators. For instance, annotators could pick a different level of hierarchy (e.g. *Fatigue* or more specific term *Tiredness*) or inconsistently pick from *similarly described* concepts when a description is vague (e.g. *Insomnia* and *Poor quality sleep*). As a result, such variable annotations cannot be simply merged and utilised, and therefore, such data become less valuable when developing machine learning-based concept normalisation systems. To combine and harmonise datasets, we need to tackle various problems associated with providing cross-terminology mappings between concepts and resolving inconsistent annotations from different datasets. Due to heterogeneous structures of medical terminologies, simple one-to-one mappings may be insufficient to match and compare concepts. Therefore, it is also necessary to harmonise and align terminologies and find a way to represent medical concepts considering relations between them regardless of the terminology. Representation learning techniques have shown promising results in encoding structural information about nodes in graphs and heterogeneous networks (Perozzi et al., 2014; Grover and Leskovec, 2016; Dong et al., 2017; Hamilton et al., 2017), however this requires integrating various medical terminologies into a single graph or network, which remains challenging. Recently, it has been also demonstrated that terminological embeddings can capture semantic similarities and are especially well-suited for biomedical ontology alignment (Kolyvakis et al., 2018). In this paper, we present a MedNorm corpus consisting of 27,979 textual descriptions (phrases) simultaneously mapped to both MedDRA and SNOMED-CT, that have been sourced from five publicly available datasets across biomedical and social media domains. To combine them, we designed a data harmonisation pipeline that can be re-used in the future to integrate new datasets into the corpus or applied in relevant annotation and data processing tasks. Also, we have described a method to merge multiple medical terminologies into a single network preserving both terminology-specific and cross-terminology relations. We demonstrated that representation learning approach based on random walks on a graph can efficiently encode equivalent and hierarchical relations and capture semantic similarities not only between concepts inside a given terminology, but also between concepts from different terminologies. Finally, we have provided an analysis of the corpus, investigated textual and conceptual similarities between utilised datasets and also analysed cross-terminology medical concept embeddings. The corpus and concept embeddings[1] as well as the harmonisation pipeline[2] are publicly available. Making such resources available to the research community aimed to contribute to a better understanding of the task.

2 Corpus material

2.1 Target medical ontologies

Relationships between medical concepts are encoded differently in medical ontologies. In this section we describe the two ontologies that have been used for mappings in the corpus.

SNOMED-CT (SCT) is a structured clinical terminology that enables consistent documentation and annotation of clinical data. There are both hierarchical and semantic (e.g. finding site, associated morphology) relations between terms. Each term can have multiple hierarchical paths with different lengths, so their *specific level* in the hierarchy is undefined.

MedDRA is a hierarchical terminology with five levels (from very specific to very general) designed for encoding adverse drug events for regulatory affairs. The most specific level is Lowest Level Terms (LLT) and refers how a concept might be reported in practice (e.g. *"Feeling queasy"*). Each LLT is linked to exactly one Preferred Term (PT), a distinct descriptor for a symptom, sign, disease diagnosis, indication, procedure or medical history characteristic (e.g. *"Nausea"*). Related PTs are grouped into High Level Terms (HLTs, e.g. *"Nausea and vomiting symptoms"*), then into High Level Group Terms (HLGTs, e.g. *"Gastrointestinal signs and symptoms"*), and finally into "System Organ Classes" (SOC, e.g. *"Gastrointestinal disorders"*). Note that single HLT can be linked to more than one HLGTs, and as a result, PT will have more than one hierarchical path to SOC.

2.2 Source corpora

The data for the MedNorm corpus was collected across two different domains: *biomedical documents* (drug labels and PubMed abstracts) and *social media* (online health forums and drug-related discussions in Twitter). The list of source datasets and their descriptions are provided below. Table 1 represents the overview of utilised terminologies.

Dataset	UMLS	MedDRA	SCT
CADEC	✗	✓*	✓
TwADR-L	✓	✗	✗
TwiMed	✓	✗	✗
SMM4H-2017	✗	✓	✗
TAC 2017 (ADR)	✗	✓	✗

* - partially mapped to MedDRA (only ADR mentions)

Table 1: Terminologies used in publicly available datasets to annotate medical concepts.

CADEC: The CSIRO Adverse Drug Event Corpus (CADEC) (Karimi et al., 2015) is an annotated corpus of patient-reported adverse drug events (ADEs) sourced from the medical forum called AskAPatient[3], which collects ratings and reviews of medications from their consumers. It contains

---

[1]https://dx.doi.org/10.17632/b9x7xxb9sz.1

[2]https://github.com/mbelousov/MedNorm-corpus

[3]https://www.askapatient.com

1,250 forum posts annotated for mentions of *Drug*, *ADR*, *Disease*, *Symptom* and *Finding*. Every mention other than *Drug* has been mapped to the corresponding SNOMED-CT concept identifier, whereas ADR mentions have been also mapped to the corresponding MedDRA term.

TwADR-L: The `TwADR-L` dataset has been constructed by the University of Cambridge (Limsopatham and Collier, 2016) from a collection of three months of Twitter posts, which has been sampled and annotated by undergrad-level linguists who mapped each phrase to one of the concepts in the UMLS Metathesaurus.

TwiMed: A corpus consists of 1,000 tweets and 1,000 PubMed sentences selected using the same strategy and annotated by two pharmacists for a set of drugs, diseases and symptoms (Alvaro et al., 2017). The `TwiMed-Twitter` set contains 827 phrases and the `TwiMed-PubMed` contains 1,142 phrases, both mapped to the UMLS Metathesaurus.

SMM4H-2017: This is a dataset of concept mentions and their corresponding human-assigned MedDRA PTs has been provided as a part of the 2nd Social Media Mining for Health Applications Shared Task at AMIA 2017 (Subtask 3) (Sarker et al., 2018). It consist of two sets: the `SMM4H2017-train` set (6,650 phrases) and the `SMM4H2017-test` set (2,500 phrases).

TAC 2017 (ADR Track): The Text Analysis Conference (TAC) 2017 Shared Task had a track on Adverse Drug Reaction Extraction from Drug Labels (Demner-Fushman et al., 2018), the final task of which was focused on mapping extracted ADRs in a Structured Product Labels (SPL) to MedDRA PTs. The training set (`TAC2017_ADR`) of 101 annotated drug labels has been released, which contain 7,045 ADR mentions mapped to MedDRA.

3 Corpus creation

The overview of the data harmonisation pipeline used to create a corpus is illustrated in Figure 1. Initially, we have combined all seven datasets from five data sources mentioned above into a single set of instances where each phrase is associated with corresponding original identifiers in different terminologies. We have represented the corpus as a graph to preserve relations between datasets and their annotations (Section 3.1). Then, we extracted hierarchical relations and linked all concepts to their closely matched (equivalent) concepts across terminologies (Section 3.2). We have encoded both hierarchical and equivalent relations between concepts in different terminologies in a low-dimensional vector space that enables to measure the similarity between them (Section 3.3). In addition, we attempted to identify and resolve potential inconsistencies in human annotations (Section 3.4). In order to achieve consistent hierarchy levels across annotations, all instances have been simultaneously mapped to either the Preferred Term (PT) or higher level (e.g. when original annotation was less specific) in MedDRA and its equivalent level in SNOMED-CT. After such process, each phrase could have more than one equivalent mapping candidate (*multi-label*). Therefore, to provide one-to-one mapping between phrases and concepts, multiple candidates have been reduced to a single concept (*single-label*). As a result, we constructed our corpus of 27,979 textual descriptions (phrases) simultaneously mapped to both MedDRA (version 21.1) and SNOMED-CT (version 2018-07-31).

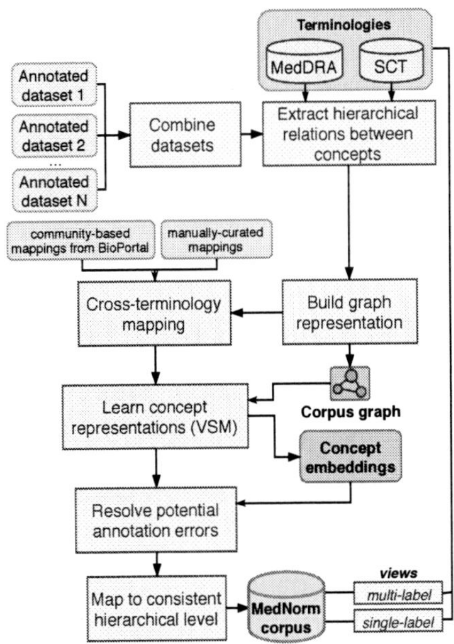

Figure 1: The data harmonisation pipeline

3.1 Building a corpus graph

In order to utilise the structure and relations of annotations in different datasets, the directed graph or network has been created (Figure 2). In such graph, each `DATASET` (e.g. *CADEC*) has a set

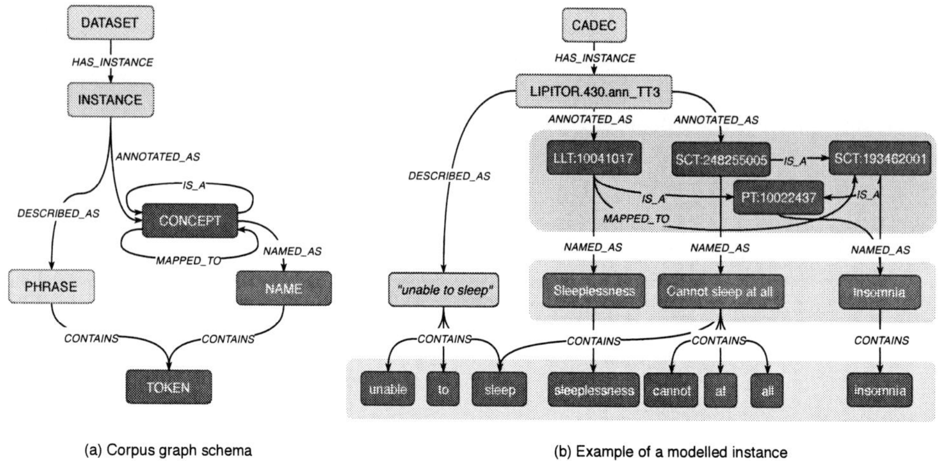

(a) Corpus graph schema   (b) Example of a modelled instance

Figure 2: Corpus graph schema (left) and an example of a modelled instance (right).

of instances, each INSTANCE can be originally annotated with one or multiple CONCEPTs (e.g. LLT:10041017, SCT:248255005) and described with textual PHRASE (e.g. *"unable to sleep"*), which in its turn contains a set of TOKENs (e.g. {*sleep, unable, to*}). Each of the CONCEPT has a corresponding NAME in the terminology (e.g. *Sleeplessness, Cannot sleep at all*), which is encoded using *NAMED_AS* link and also contains a set of tokens (similar to phrase). To represent hierarchical relations between concepts extracted from medical terminologies, each CONCEPT can be linked to its parent node (i.e. concept from the higher level in the hierarchy) with *IS_A* link (e.g. *Sleeplessness → Insomnia → Disturbances in initiating and maintaining sleep → Sleep disorders and disturbances → Psychiatric disorders*) and mapped to the equivalent concept node using *MAPPED_TO* relation (e.g. *Sleeplessness* LLT:10041017 *→ Insomnia* SCT:193462001). The representation of the corpus as a graph makes the further processing and analysis easier. For example, testing whether a particular phrase has been inconsistently annotated in the same dataset (i.e. has more than one associated concept) could be done by counting the number of unique CONCEPT nodes reachable from the target phrase. Moreover, all links between concepts in different terminologies (despite their various structures) are stored inside the single graph.

3.2 Cross-terminology mapping

The automatic mapping between UMLS, Med-DRA and SNOMED-CT has been done using community-based mappings from BioPortal (Noy et al., 2008) through the REST API [4]. The two concepts from different ontologies are considered as *equivalent* or *closely matched* if they share the same UMLS Concept Unique Identifier (CUI). After a careful review of results, we observed that some of the frequently mentioned concepts have not been mapped automatically. Therefore, with the help of medical experts, we defined an additional set of manually-curated mapping rules (provided in Appendix A, Table 6).

3.3 Learning cross-terminology representations of concepts

Cross-terminology mappings allowed to link concepts from multiple terminologies together, but their heterogeneous hierarchical structures (i.e. concepts are located deeper in the hierarchy or have more relations) makes graph distance alone insufficient to measure the similarity between concepts in different terminologies. However, medical concepts (or their corresponding nodes) can be embedded into a low-dimensional vector space. Initially, we have constructed a simplified hierarchical *concept graph* whose vertices are *groups* of equivalent concepts (i.e. nodes linked with *MAPPED_TO* relation in the main corpus graph) and edges are hierarchical *IS_A* relations. Then, we have used the DeepWalk (Perozzi et al., 2014), a deep learning method based on generalisation of language modelling applied on the streams of short random walks treating them as the equiva-

---

[4]http://data.bioontology.org

lent of sentences. Performing 10 random walks per node (with a length of 40 nodes) and training a Skip-gram model (Mikolov et al., 2013) with the window size of 5, we have generated 64-dimensional concept vectors. The size of vectors has been chosen empirically. Later, we have split the groups under the assumption that all concepts in a group (i.e. equivalent concepts) should have the same vectors. Table 2 shows three selected MedDRA concepts and their most similar concepts (with cosine similarity) from all terminologies. It demonstrates that both equivalent and hierarchical relations between concepts has been successfully encoded and the semantic similarity can be captured by calculating the cosine similarity between two corresponding concept vectors.

3.4 Corpus consistency

In order to make all annotations in our final corpus consistent, we have performed the two operations described below.

Resolving inconsistent annotations: After performing a manual analysis of the combined corpus we have noticed inconsistencies in the original human annotations. For example, in the CADEC, where phrases can be mapped simultaneously to both SNOMED-CT and MedDRA, 27 instances which were (correctly) annotated as *Stomach cramps* (SCT:51197009) also were co-annotated as *Learning disorder* (MEDDRA_PT:10061265). To identify potential annotation errors in the original datasets, we have utilised the concept graph to calculate the distances between concept nodes (i.e. the shortest path length) and the cosine similarity of corresponding vectors in the latent vector space model (VSM). Also, we made an effort to locate inconsistent annotations across different datasets by identifying *ambiguous tokens*. In the usual case, a *specific token* is used to describe groups (clusters) of similar concepts (e.g. *"walk"* frequently describes concepts related to walking or mobility). However, an *ambiguous token* describes clusters of similar concepts frequently, but also sometimes describes concepts that are different from those clusters (i.e. the difference between the number of occurrences in the groups is high). Note that *common tokens* (e.g. *"unable"*), that are not specific for a particular group of concepts, will usually have a high number of groups, but relatively small difference between the numbers of occurrences.

We attempted to identify such outliers by calculating distances between concepts and their distance deviations from the clusters. For example, token *"walk"* was mentioned in 98 phrases and mapped to 23 concepts in total. The most popular annotation was *Walking disability* (e.g. *"can barely walk"*), however it also has been annotated as *Myocardial infarction* (e.g. *"walk a little funny"*) that could be a potential annotation error. After such analysis and manual review, we have identified and re-mapped 110 annotations (provided with the source code).

Consistent hierarchical mapping: The Preferred Term (PT) level in MedDRA describes single medical concept. Therefore it has been selected as a standard to provide a consistent hierarchical level among annotations in our corpus. However, not all phrases are specific enough to be mapped to the PT level or its equivalent. In such cases, we kept annotations equivalent to higher MedDRA levels (i.e. HLT, HLGT or SOC). All lower level annotations (i.e. LLT-equivalent) have been mapped to their PT-equivalent parents. Using the corpus graph, we were able to automate this process. Initially, all instances regardless of the terminology used in original annotations have been *recursively* mapped to their corresponding equivalent PT candidates (i.e. including mappings of mappings). Then, for each MedDRA candidate, we selected equivalent candidates from SNOMED-CT. To filter concepts that have emerged from such automatic mapping, all concepts that have not been observed in the original annotations were removed (except cases, where it was the only possible candidate). After such process, each phrase could have more than one candidate for each terminology (multi-label). Therefore, to provide one-to-one mapping between phrases and terminologies, in each multi-label group we have initially identified the most similar MedDRA concept to the original annotation (i.e. from the source dataset) but also the most popular across the whole corpus (i.e. to minimise the number of outliers). Then, we selected the SNOMED-CT concept (from the multi-label group) that is the most similar to the selected MedDRA concept to achieve consistency in mapping between terminologies. Hereby, each phrase has been mapped to exactly one (single-label) MedDRA and its corresponding SNOMED-CT concept simultaneously. As a result, the final corpus

Insomnia PT:10022437	Weight increased PT:10047899	Nausea PT:10028813
1.0000 Insomnia disorder LLT:10078083	1.0000 Ponderal increased LLT:10063441	1.0000 Nauseous LLT:10028823
1.0000 Insomnia SCT:193462001	1.0000 Wt gain LLT:10048060	1.0000 Feeling queasy LLT:10016361
1.0000 Insomnia NOS LLT:10022442	1.0000 Weight increasing SCT:161831008	1.0000 Nauseated LLT:10028822
1.0000 Sleeplessness LLT:10041017	1.0000 Weight increase LLT:10047898	1.0000 Nausea SCT:422587007
1.0000 Sleeplessness C0917801	1.0000 Weight gain finding SCT:8943002	1.0000 Nausea C0027497
0.9795 Sleep loss C0235161	1.0000 Weight gain C0043094	1.0000 Queasy LLT:10037730
0.9795 Sleep loss LLT:10041001	1.0000 Weight increased SCT:262286000	0.8677 Nausea and vomiting symptoms HLT:10028817
0.9795 Sleep decreased LLT:10040982	1.0000 Weight gain LLT:10047896	0.8677 Nausea and vomiting SCT:16932000
0.9779 Middle insomnia SCT:67233009	0.9532 Weight change finding SCT:365921005	0.8677 Nausea and vomiting C0027498
0.9779 Middle insomnia C0393761	0.9532 Weight change finding C1287464	0.7902 Gastrointestinal tract finding C1261141
0.9779 Sleep maintenance insomnia LLT:10068671	0.9375 Weight loss finding SCT:89362005	0.7902 Gastrointestinal tract finding SCT:386618008
0.9779 Middle insomnia PT:10027590	0.9375 Weight decreased PT:10047895	0.7832 Travel sickness NOS LLT:10044549
0.9743 Trouble falling asleep LLT:10044698	0.9375 Weight decreased SCT:262285001	0.7832 Motion sickness C0026603
0.9743 Initial insomnia C0393760	0.9375 Wt loss LLT:10048061	0.7832 Travel sickness LLT:10044548
0.9743 Initial insomnia SCT:59050008	0.9375 Weight decreasing SCT:161832001	0.7832 Motion sickness PT:10027990
0.9743 Initial insomnia PT:10022035	0.9375 Lost weight LLT:10024886	0.7832 Motion sickness SCT:37031009
0.9689 Early morning awakening LLT:10014046	0.9375 Loss of weight LLT:10024883	0.7721 Retching C0232602
0.9689 Terminal insomnia PT:10068932	0.9375 Weight decrease LLT:10047893	0.7721 Dry heaves LLT:10052104
0.9689 Terminal insomnia SCT:67062000	0.9375 Losing wt LLT:10024849	0.7721 Retching SCT:84480002
0.9689 Awakening early LLT:10003867	0.9375 Weight loss LLT:10047900	0.7721 Vomiturition LLT:10072124

Prefixes for concept identifiers: SCT - SNOMED-CT; C - UMLS; LLT, PT, HLT, HLGT, SOC - MedDRA (based on the level). The equivalent concepts have similarity value of 1.0.

Table 2: MedDRA concepts and their most similar concepts across different terminologies.

Phrase	Original annotations	Mapped MedDRA	Mapped SNOMED-CT
screwed my endocrine system	Endocrine disorders SOC:10014698	***Endocrine disorders SOC:10014698*** Endocrine disorder PT:10014695	***Disorder of endocrine system SCT:362969004***
Got 1.5 hours of sleep	Sleep disturbance C0037317	***Sleep disturbances HLGT:10040998*** Sleep disorder PT:10040984	***Disturbance in sleep behavior SCT:53888004*** Sleep disorder SCT:39898005
wrecking my sleep	Poor quality sleep C1262141	Poor quality sleep PT:10062519 Dyssomnia PT:10061827 ***Sleep disorder PT:10040984***	Dyssomnia SCT:44186003 ***Sleep disorder SCT:39898005***
all I want to do is sleep	Somnolence PT:10041349	Somnolence PT:10041349 ***Insomnia PT:10022437***	Drowsy SCT:271782001 ***Insomnia SCT:193462001***
weak	Asthenia PT:10003549	***Asthenia PT:10003549***	***Asthenia SCT:13791008***
fatigue	Fatigue C0015672	***Fatigue PT:10016256*** Asthenia PT:10003549	***Fatigue SCT:84229001*** Asthenia SCT:13791008 Lack of energy SCT:248274002
extremely tired feeling	Tiredness LLT:10043890 Feeling tired SCT:314109004	***Fatigue PT:10016256*** Asthenia PT:10003549	***Fatigue SCT:84229001*** Asthenia SCT:13791008 Lack of energy SCT:248274002 Feeling tired SCT:249274002

Selected concepts (during multi-label reduction to single-label) are in ***bold-italic***.

Table 3: Examples of originally annotated phrases and their multi-label and single-label mappings

has 27,957 PT-equivalent, two HLT-equivalent, 18 HLGT-equivalent and two SOC-equivalent annotations. In Table 3 we have provided examples of phrases, original annotations and our final Med-DRA and SNOMED CT annotations (mappings).

4 Corpus analysis

The descriptive statistics of datasets constituting a corpus (grouped into *biomedical* and *social* domains) are presented in Table 4. The length of medical concept descriptions (phrases) are longer in social domain. The longest phrase has been found in the CADEC corpus: *"when I went to sit down instead of siting normally I would almost fall down in the chair no control no strength, upon getting up I had to hold on to something to get up"*

(36 tokens) that describes *Muscle weakness*. We have also investigated the degree of class imbalance in the corpus and illustrated the most reported MedDRA concepts in Figure 3. The most reported

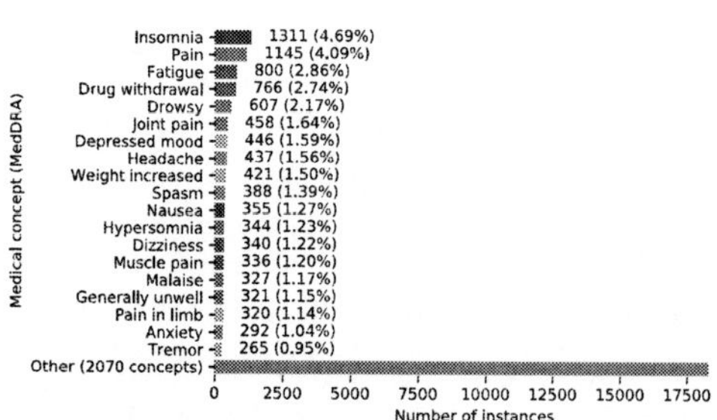

Figure 3: Most popular concepts in the corpus

Dataset	# inst	# MedDRA*	# SCT*	# phrases	# words	phrase length
TAC2017_ADR	5,835	1,113	1,087	2,106	1,633	$2.46 \pm 1.49\ [1-13]$
TwiMed-PubMed	1,067	254	255	436	478	$1.93 \pm 1.11\ [1-8]$
All biomedical	6,902	1,191	1,169	2,397	1,804	$2.42 \pm 1.46\ [1-13]$
CADEC	6,797	530	557	3,376	1,966	$3.42 \pm 2.26\ [1-36]$
SMM4H2017-train	6,416	411	404	2,638	2,084	$3.24 \pm 2.22\ [1-25]$
TwADR-L	4,626	1544	1566	2,581	2,492	$2.46 \pm 1.78\ [1-20]$
SMM4H2017-test	2,447	227	224	1,148	1,165	$3.31 \pm 2.33\ [1-18]$
TwiMed-Twitter	791	185	185	428	524	$2.08 \pm 1.49\ [1-12]$
All social	21,077	1,740	1,778	8,890	4,975	$3.26 \pm 2.21\ [1-36]$
ALL	27,979	2,062	2,089	10,572	5,584	$3.18 \pm 2.12\ [1-36]$

* - single-label annotations

Table 4: Statistics of the datasets constituting the corpus.

concept is *Insomnia* (1,311 instances, 553 unique phrases), followed by *Pain* (1,145 instances, 320 unique phrases) and *Fatigue* (800 instances, 125 unique phrases). However, about 40% of concepts were under-reported and have only one instance, corresponding to about 3% instances in the whole corpus. The average number of unique phrases per terminology concept is 5.13 for MedDRA and 5.06 for SNOMED-CT.

4.1 Asymmetric transferability between datasets

To investigate how the knowledge acquired from one dataset is potentially transferable to another dataset, we introduced the *asymmetric transferability index* that takes into account both *conceptual* (i.e. concepts from various terminologies used in the dataset) and *textual* (i.e. language used to describe those concepts) similarities. Asymmetry allows to see how much information can be understood from another dataset having all information about the first dataset. It utilises two similarity measures: cosine similarity $CS(X, Y) = \frac{X \cdot Y}{\|X\|\|Y\|}$ and the special case of Tversky Index (Tversky, 1977) with $\alpha = 1$ and $\beta = 0$, that can be rewritten as $TI(X, Y) = \frac{|X \cap Y|}{|X \cap Y| + |X - Y|}$. We can calculate the similarity between two sequences of labels l_1 and l_2 with the cosine similarity between the corresponding label count vectors $c(l_1)$ and $c(l_2)$. However that measure will be symmetric, and therefore we multiply it by asymmetric set-based similarity:

$$s(l_1, l_2) = TI(l_1, l_2) \times CS(c(l_1), c(l_2)) \quad (1)$$

Having two datasets A and B, sets of phrases P_A, P_B and sets of words W_A, W_B we obtain the *textual transferability index* (from A to B) as the arithmetic mean of phrasal and verbal asymmetric similarities:

$$I_{txt}(A, B) = \frac{TI(P_A, P_B) + TI(W_A, W_B)}{2} \quad (2)$$

For each terminology t, we extract sequences of labels $\ell(A, t)$ in dataset A and $\ell(B, t)$ in dataset B. The *conceptual transferability index* is the average asymmetric similarity between terminology-specific label sets:

$$I_{con}(A, B) = \frac{1}{|T|} \sum_{t \in T} s(\ell(A, t), \ell(B, t)) \quad (3)$$

Finally, we obtain the *overall transferability index*:

$$I_{ovr}(A, B) = \frac{I_{txt}(A, B) + I_{con}(A, B)}{2} \quad (4)$$

We have presented textual, conceptual and overall transferability matrices in Figure 4. The higher transferability index shows the better chance to understand information (i.e. match vocabulary or concepts). The most transferable dataset was TwADR-L, whereas the least transferable was TwiMed-PubMed. It directly corresponds to the number of unique concepts, phrases and words reported previously in Table 4. Also, the datasets collected from Twitter are highly transferable between each other. The CADEC dataset collected from AskAPatient reports is still more similar to Twitter (i.e. social domain).

4.2 Cross-terminology concept representations

In order to analyse cross-terminology concept representations, we used T-distributed Stochastic Neighbour Embedding (t-SNE) (Maaten and Hinton, 2008) to perform dimensionality reduction from 64D to 2D (Figure 5). It can be observed that semantically similar concepts have been clustered together, providing additional evidence about the

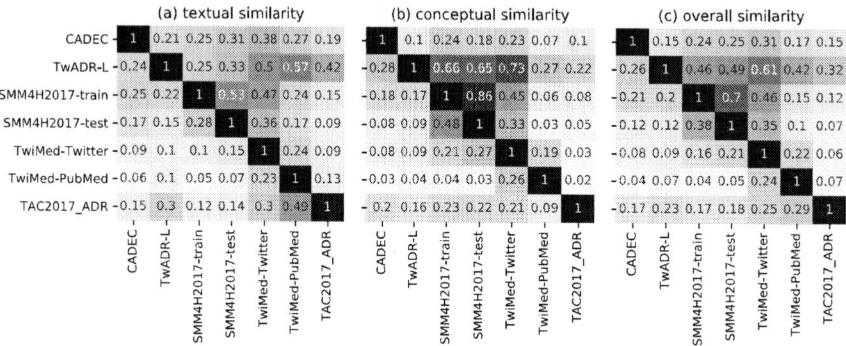

Figure 4: Asymmetric dataset transferability matrices.

ability of concept representations to encode hierarchical and equivalent relations and capture semantic similarities.

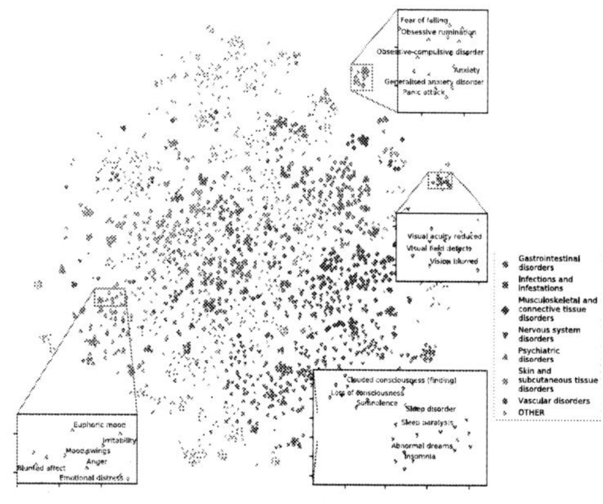

Figure 5: t-SNE visualisation of cross-terminology medical concept representations

In Table 5 we have presented the most similar MedDRA and SNOMED-CT annotations (i.e. the final labels in the corpus) for the three most frequently reported concepts: Insomnia, Pain and Fatigue. Although such representations encoded conceptual similarity well, they are insufficient to identify *opposite* concepts correctly (e.g. *Fatigue* and *Energy increased*). This is because we only utilised hierarchical relations in terminologies (information about opposite concepts is not provided in these terminologies explicitly).

5 Conclusion

We have presented a corpus for cross-terminology medical concept normalisation that has been ·sourced from five publicly available datasets across the biomedical and social domains. The

Concept	MedDRA	SNOMED-CT
Insomnia	0.98 Middle insomnia	0.98 Middle insomnia
	0.97 Initial insomnia	0.97 Initial insomnia
	0.97 Terminal insomnia	0.97 Early morning waking
	0.97 Hyposomnia	0.97 Not getting enough sleep
	0.91 Poor quality sleep	0.91 Dyssomnia
Pain	0.82 Labour pain	0.82 Labor pain
	0.78 Nyctalgia	0.78 Night pain
	0.76 Tenderness	0.76 Tenderness
	0.60 Painful respiration	0.68 Burning epigastric pain
	0.58 Odynophagia	0.68 Postoperative pain
Fatigue	0.83 Asthenia	0.83 Asthenia
	0.83 Lethargy	0.83 Lethargy
	0.69 Malaise	0.77 Sensation of heaviness in limbs
	0.69 Feeling abnormal	0.69 Generally unwell
	0.68 Energy increased	0.69 Malaise

Table 5: Most similar MedDRA and SNOMED-CT concepts (from annotations).

data harmonisation pipeline described in the paper combines instances from various datasets and provides consistent simultaneous mappings to both MedDRA and SNOMED-CT terminologies. Such pipeline can be used in the future to integrate new datasets into the corpus or could be also applied in relevant data annotation and processing tasks. Also, we have described a method to merge multiple medical terminologies and demonstrated that equivalent and hierarchical relations can be encoded into cross-terminology concept representations that are able to capture semantic similarities not only between concepts inside a given terminology but also between concepts from different terminologies. The generated cross-terminology medical concept representations can be used to improve and analyse the performance of concept normalisation systems. Making such resources available to the research community as well as providing an analysis of the final corpus aimed to contribute to a better understanding of the task and associated challenges.

References

Nestor Alvaro, Yusuke Miyao, and Nigel Collier. 2017. Twimed: Twitter and pubmed comparable corpus of drugs, diseases, symptoms, and their relations. *JMIR public health and surveillance*, 3(2).

Elliot G Brown, Louise Wood, and Sue Wood. 1999. The medical dictionary for regulatory activities (meddra). *Drug safety*, 20(2):109–117.

Dina Demner-Fushman, Sonya E Shooshan, Laritza Rodriguez, Alan R Aronson, Francois Lang, Willie Rogers, Kirk Roberts, and Joseph Tonning. 2018. A dataset of 200 structuerred product labels annotated for adverse drug reactions. *Scientific data*, 5:180001.

Yuxiao Dong, Nitesh V Chawla, and Ananthram Swami. 2017. metapath2vec: Scalable representation learning for heterogeneous networks. In *Proceedings of the 23rd ACM SIGKDD international conference on knowledge discovery and data mining*, pages 135–144. ACM.

Aditya Grover and Jure Leskovec. 2016. node2vec: Scalable feature learning for networks. In *Proceedings of the 22nd ACM SIGKDD international conference on Knowledge discovery and data mining*, pages 855–864. ACM.

William L Hamilton, Rex Ying, and Jure Leskovec. 2017. Representation learning on graphs: Methods and applications. *arXiv preprint arXiv:1709.05584*.

Sarvnaz Karimi, Alejandro Metke-Jimenez, Madonna Kemp, and Chen Wang. 2015. Cadec: A corpus of adverse drug event annotations. *Journal of biomedical informatics*, 55:73–81.

Prodromos Kolyvakis, Alexandros Kalousis, Barry Smith, and Dimitris Kiritsis. 2018. Biomedical ontology alignment: an approach based on representation learning. *Journal of biomedical semantics*, 9(1):21.

Nut Limsopatham and Nigel Henry Collier. 2016. Normalising medical concepts in social media texts by learning semantic representation. Association for Computational Linguistics.

Laurens van der Maaten and Geoffrey Hinton. 2008. Visualizing data using t-sne. *Journal of machine learning research*, 9(Nov):2579–2605.

Tomas Mikolov, Kai Chen, Greg Corrado, and Jeffrey Dean. 2013. Efficient estimation of word representations in vector space. *arXiv preprint arXiv:1301.3781*.

Natalya F Noy, Nicholas Griffith, and Mark A Musen. 2008. Collecting community-based mappings in an ontology repository. In *International Semantic Web Conference*, pages 371–386. Springer.

Bryan Perozzi, Rami Al-Rfou, and Steven Skiena. 2014. Deepwalk: Online learning of social representations. In *Proceedings of the 20th ACM SIGKDD international conference on Knowledge discovery and data mining*, pages 701–710. ACM.

Abeed Sarker, Maksim Belousov, Jasper Friedrichs, Kai Hakala, Svetlana Kiritchenko, Farrokh Mehryary, Sifei Han, Tung Tran, Anthony Rios, Ramakanth Kavuluru, et al. 2018. Data and systems for medication-related text classification and concept normalization from twitter: insights from the social media mining for health (smm4h)-2017 shared task. *Journal of the American Medical Informatics Association*, 25(10):1274–1283.

Peri L Schuyler, William T Hole, Mark S Tuttle, and David D Sherertz. 1993. The umls metathesaurus: representing different views of biomedical concepts. *Bulletin of the Medical Library Association*, 81(2):217.

Michael Q Stearns, Colin Price, Kent A Spackman, and Amy Y Wang. 2001. Snomed clinical terms: overview of the development process and project status. In *Proceedings of the AMIA Symposium*, page 662. American Medical Informatics Association.

Amos Tversky. 1977. Features of similarity. *Psychological review*, 84(4):327.

A   Appendices

MedDRA Concept	SNOMED-CT Concept
Withdrawal syndrome (PT:10048010)	Drug withdrawal (SCT:363101005)
Depression (PT:10012378)	Depressive disorder (SCT:35489007)
Drug ineffective (PT:10013709)	Lack of drug action (SCT:58848006)
Hangover (PT:10019133)	Hangover (SCT:32553006)
Infection (PT:10021789)	Infectious disease (SCT:40733004)
Feeling abnormal (PT:10016322)	Malaise (SCT:367391008)
Feeling jittery (PT:10016338)	Feeling nervous (SCT:424196004)
Poor quality sleep (PT:10062519)	Dyssomnia (SCT:44186003)
Thirst (PT:10043458)	Thirst symptom (SCT:249475006)
Lightheadedness (LLT:10024492)	Lightheadedness (SCT:386705008)

Table 6: An additional set of manually-curated mapping rules between MedDRA and SNOMED-CT.

Passive Diagnosis incorporating the PHQ-4 for Depression and Anxiety

Fionn Delahunty[1,2] **Robert Johansson**[3,4] **Mihael Arcan**[2]

[1] University of Gothenburg | Chalmers University of Technology, Sweden
[2] Insight Centre for Data Analytics, Data Science Institute, NUI Galway, Ireland
[3] Department of Psychology, Stockholm University, Sweden
[4] Department of Computer and Information Science, Linköping University, Sweden
`{Fionn.Delahunty,Mihael.Arcan}@Insight-centre.org`
`Robert.Johansson@liu.se`

Abstract

Depression and anxiety are the two most prevalent mental health disorders worldwide, impacting the lives of millions of people each year. In this work, we develop and evaluate a multilabel, multidimensional deep neural network designed to predict PHQ-4 scores based on individuals written text. Our system outperforms random baseline metrics and provides a novel approach to how we can predict psychometric scores from written text. Additionally, we explore how this architecture can be applied to analyse social media data.

1 Introduction

According to the World Health Organization (WHO), major depressive disorder[1] is the largest cause of disability worldwide (World Health Organization, 2018), with a lifetime prevalence rate between 15% and 17% (Ebmeier et al., 2006). Depression is highly co-morbid with several other mental disorders, the most prevalent of which is a generalized anxiety disorder.[2] Almost 50% of individuals diagnosed with depression will also be diagnosed with anxiety (Johansson et al., 2013).

As a result, many clinicians will investigate for the presence of both disorders at the time of diagnosis. To do so, psychometric questionnaires are often employed as a quick and reliable initial assessment tool, the most common of which is the Patient Health Questionnaire (PHQ). The PHQ-4 is a short form questionnaire design to access the presence or absence of the core symptoms in depression and anxiety (Löwe et al., 2010). The questionnaire has demonstrated both high validity and reliability across several languages and cultures (Kroenke et al., 2010).

Despite the usefulness of these questionnaires, there is still a reliance on individuals actively seeking a diagnosis from a medical professional before they can be applied. Research has shown that those suffering depression and anxiety often are unaware their symptoms are due to a medical disorder and attribute them to poor mood or external factors (Barney et al., 2006; Latalova et al., 2014). This presents a unique challenge in the medical community, in how to inform and encourage individuals to come forward for diagnosis.

Delahunty et al. (2018) have proposed the concept of **passive diagnosis**, also known as high-performance medicine (Topol, 2019). This term refers to the ability for machine learning algorithms to constantly monitor an individuals health and inform the individual if certain changes are evidence of a possible disorder in the future. This is in comparison to the traditional concept of active diagnosis where an individual suffering certain symptoms would actively seek out a medical diagnosis.

Examples of applications in this domain include *DeepCare*, which is an end-to-end application designed to diagnose a wide range of disorders (Pham et al., 2016). Such systems allow clinicians to either prevent a disorder occurring or provide early intervention to minimise its effects.

2 Related work

While exploring the effects of expressive writing on PTSD[3] treatment, (Pennebaker et al., 2003) established that the way in which individuals wrote was often indicative of their mental state, specifically their use of function words (Prendinger and Ishizuka, 2005). Examples of this included higher counts of the personal pronouns and negative

---

[1]Hereafter referred to as simply depression.
[2]Hereafter referred to as simply anxiety.

[3]Post-Traumatic Stress Disorder

Proceedings of the 4th Social Media Mining for Health Applications (#SMM4H) Workshop & Shared Task, pages 40–46
Florence, Italy, August 2, 2019. ©2019 Association for Computational Linguistics

words in depressed individuals' writing, which is attributed to a manifestation of Beck's cognitive model and Pyczsinski and Greenberg's self-focus model of depression (Rude et al., 2004).

Over recent years this work has been combined with the fields of natural language processing and machine learning to develop classifiers algorithms which can predict if an individual is likely to be diagnosed with a certain disorder. Work has focused on bipolar disorder (Huang et al., 2017), depression (De Choudhury et al., 2013) and anorexia (Ramiandrisoa and Benamara, 2018). For the last number of years, the *CLEF* conference has hosted a workshop on early risk prediction of mental disorders based on social media data (Losada and Crestani, 2016), resulting in almost 50 publications in this area.

However, much of the existing work suffers from the limitation of viewing these disorders as binary occurrences, whether a disorder is present or not. Although this approach makes sense given the nature of machine learning classifiers, from the perspective of medical professionals, however, individuals can rarely be placed into binary classes. Different combinations of symptoms can dramatically affect the diagnosis (American Psychiatric Association, 2013).

Previous work in 2018 was the first to view these disorders on a symptomatic level (Delahunty et al., 2018). In this paper, we expand the previous work by including anxiety and making use of the PHQ, which compared to the Beck's depression inventory is a non-commercial psychometric questionnaire (Kung et al., 2013).

The PHQ-4 assess the severity of the two primary symptoms for depression and anxiety respectfully, anhedonia, depressed mood, excessive anxiety and uncontrollable worry (American Psychiatric Association, 2013). An individual is asked to rate the occurrence of each symptom over the last two weeks on a four-point scale from "Not at all" to "Nearly every day". The aim of our work was to develop a machine learning algorithm that given textual data could predict an outcome value for each of the four questions on the PHQ-4. Unlike previous work, e.g. Delahunty et al. (2018), we did not employ separate algorithms for the four symptoms, but considered that all four symptoms are intrinsically interconnected. Within the machine learning literate, multilabel and multi-class approaches have been shown to outperform indi-

vidual separate classifiers (Schmidhuber, 2015).

Previous work in this domain has often employed extracted data from social media sites as training data (Losada and Crestani, 2016). In many cases, this limits the application of the work because it is impossible if the individuals in the training data actually had clinical diagnosis (De Choudhury and De, 2014). To overcome this limitation, our work employs a dataset collected in an in-person medical setting where clinical diagnosis are performed by trained professionals. We aim to explore if training on non-social media data will allow for accurate evaluation on social media data.

3 System Description

3.1 Data

Our initial dataset is the DAIC-WOZ, which is composed of transcribed clinical interviews collected through a Wizard-of-Oz approach for 142 patients (Gratch et al., 2014). The topic of the interviews are general conversations and were all collected within the United States. For each patient, a transcript of their interview is provided along with PHQ-8 scores, where bot statements were removed leaving only patient statements. PHQ-8 scores can be mapped to PHQ-2 scores, and GAD-2 scores were inferred from data provided by Johansson et al. (2013). The final dataset was composed of 23,726 text statements.

To evaluate our system on social media data, we employed the *Reddit* depression dataset (Losada and Crestani, 2016). This dataset we gained access to contained *Reddit* posts for 253 users (of which 161 are attributed as to be suffering depression). Diagnosis is binary (depressed or not depressed) depending on if users post on certain depression sub-forums.

3.2 Feature extraction

Three methods of feature extraction were employed.
Text representation was employed using the Universal Sentence Encoder (USE), specifically developed for longer than word representations. The model is trained using a deep learning transformer neural network architecture on a variety of datasets (Cer et al., 2018). Each of our patient statements was passed into their pretrained model and a statement level representation vector of shape 512 was returned.

LIWC is a psycholinguistic dictionary containing 94 psychological trait dimensions and over 2,000 words related to these dimensions (Pennebaker et al., 2001). A percentage count of the number of words in the text related to each dimension is computed. To identify an optimal subset of the number of relevant dimensions, we reviewed all proceedings from the CLEF *eRisk* workshop 2017 and 2018 (Losada and Crestani, 2016). For each proceeding that employed LIWC, the list of dimensions included was taken. An intersection of these lists was then taken to create a subset of 22 relevant dimensions, which resulted in the following features being included in our model: *word count, analytical thinking, authentic, emotional tone, function words, pronoun, personal pronouns, 1st person singular, 1st person plural, 2nd person, 3rd person singular, articles, auxiliary verbs, conjunctions, negations, regular verbs, negative emotions, social words, cognitive processes, past focus, present focus, future focus.*

Psychometric similarity Recent work has seen success in comparing word embeddings in terms of semantic similarity (Mihalcea et al., 2006; Li et al., 2003), where the distance between embeddings in x^N-dimensional space is considered equal to their likeness in terms of the semantic content. Since USE creates sentence level embeddings, this allows us the ability to compare sentences in terms of similarity. We employed this approach by comparing the semantic similarity of patient statements with responses from psychometric questionnaires. The principle was that if a patient statement reflected the same content of a psychometric test it should have a higher similarity score compared with a random statement.

Four questionnaires were identified by choosing cognitive theories relevant to the aetiology of each of the four PHQ-4 symptoms. Details regarding the theories are included in Table 1. The concatenation of questions across all four questionnaires amounted to 104 questions. For each patient statement, a 512 embedding dimension was computed with the USE pre-trained model, along with this, embeddings for each of the 104 patient questions were computed. The inner dot product for each statement and question was computed and returned as a feature. The inner dot product measures how close two vectors are in the Euclidean space of the trained model, closer vectors implies more similar semantic similarity.

The resulting dataset was composed of 638 features. All features were scaled by removing the mean and scaling to unit variance within the bounds of -1 and 1.

3.3 Our approach

To model the interconnectivity of the four PHQ-4 symptoms, we employed a deep neural network (DNN) architecture. Unlike simpler algorithms, such as classical regression, which uses a single function, $(Y \approx f(X, \beta))$,[4] DNNs employ a large number of "neurons", each of which is fitted with an independent function with a set of weights and an activation function (Schmidhuber, 2015). Current work demonstrates that this architecture models the internal representation better than separate classifiers (Schmidhuber, 2015).

For each patient statement, the neural network needs to be able to output an ordinal value score for each question. This requires that the network outputs both multilabel (four symptoms) and multivalue (ordinal score). This architecture is regarded as multi-dimensional or multi-targeted classification, where the output is assigned both a set of labels $y = (y_0, \ldots, y_d)$, and for each label y an ordinal value in the 0 to d (Read et al., 2014). These methods are still in early development are mostly untested outside of theoretical proposals.

Our proposed method to address this problem is a two-step approach. Firstly, we apply a multilabel learning approach to constantly predict a *Sigmoid* score for each of the four symptoms. This is achieved by using a binary cross entropy loss function that can model the interconnectivity of the labels (Trotzek et al., 2018; Nam et al., 2014; Zhang and Zhou, 2014; Mencıa and Fürnkranz, 2008) and a *Sigmoid* function on the final layer (Trotzek et al., 2018). Secondly, following that, we set manual threshold values to refine this score into ordinal values for interpretability.

For the final output per symptom, we set the value to 0, if the outcome of the *Sigmoid* function is less than 0.25, 1 if the *Sigmoid* score is between 0.25 and 0.50, 2 between 0.50 and 0.75 and 3, if the *Sigmoid* score is larger than 0.75.

To compare our approach against a simpler model architecture, and determine if a DNN architecture is appropriate, we also trained a random forest classifier which is equally able to model

---

[4]Y is dependent variable, X is independent variable & β is unknown parameter.

PHQ-4 Symptom	Theory	Assessment tool
Feeling nervous or anxious	Intolerance of uncertainty (Clark et al., 1994)	Intolerance of Uncertainty Scale
Uncontrollable worry	Positive belief about worry (Clark et al., 1994)	Penn State Worry Questionnaire
Anhedonia	Avoidance behaviour (Clark et al., 1994)	Cognitive-Behavioural Avoidance Scale
Depressed mood	Negative triad (Beck, 1991)	Beck's depression inventory

Table 1: Summary of aetiology theories and assessment tools.

multilabel outputs (Gharroudi et al., 2014).

3.4 Hyperparameter Tuning

Hyperparameter tuning was achieved using a genetic algorithm approach. This approach takes its basis from the biological concept of evolution (Friedrichs and Igel, 2005). A broad set of hyperparameters are chosen (details in the appendix), the algorithm creates a generation by choosing a random subset of these and trains a population of 20 network networks with different random hyperparameters. Each network is evaluated on a metric, in this case, the minimization of the *Hamming loss* criteria (Zhang and Zhou, 2014). The five best networks based on this metric are chosen, along with five random ones to allow some variability in the population. Another generation is created with random hyperparameters chosen from within the subset of the last generation, while we repeat this process for a total of ten generations.

The final optimal hyperparameters, based on the minimized *Hamming loss*, were six dense layers with dimensions of 1024, 768, 256, 128, 64 and 4 in that order. Each layer contained a *relu* activation function, except for the final layer, which contained *Sigmoid*. Binary cross entropy was applied to compute the loss function and *adagrad* function as the optimizer.

The following hyperparameters were employed for the RFC, number of trees in the forest = 10, split criterion = gini, no max depth of trees, minimum samples to split a tree = 2, minimum leaf sample = 1.

4 Results

Multilabel Our evaluation was first performed on the multilabel aspect of the network. PHQ scores were reduced to a binary class (0 for 0, 1 for 1,2,3) and *Sigmoid* outputs were binarized on a cutoff point of 0.5. *Hamming loss* was the chosen metric for evaluation (Zhang and Zhou, 2014), which computes the distance between predicted and true values. A ten-fold cross-validation resulted in a score of 0.388, 95% [0.3870, 0.3905]. To compare this against a random baseline, where a set

Question	Accuracy	Sensitivity	Specificity
1	0.25	0.96	0.66
2	0.58	0.79	0.84
3	0.39	0.87	0.91
4	0.32	0.94	0.71

Table 2: Sensitivity and specificity scores for each question as predicted by the model

	Precision	Recall	F1-score
Depressed	0.16	0.17	0.16
Non-Depressed	0.59	0.56	0.57

Table 3: Classification scores from the *eRisk* data

of prediction scores are computed using a random number generator, a *Hamming loss* of 0.49, 95% [0.481, 0.519], is achieved.

Multidimensional Using the cutoffs mentioned above, *Sigmoid* scores were transformed into ordinal values. Since the *Hamming loss* is unsuited to this evaluation, a more suitable metric is the *Example Accuracy*, which consists of comparing if the prediction of each individual is completely correct (all values match) or incorrect and taking the mean value across all predictions (Read et al., 2014). The result across ten-fold cross-validation is 0.221, 95% [0.201, 0.243]. In comparison to ten-fold cross-validation of the RFC which resulted in a score of 0.087, 95% [0.086, 0.085].

$$EX.\ ACCURACY = \frac{1}{N} \sum_{i=1}^{N} I(\hat{y}^{(i)}, y^{(i)}) \quad (1)$$

Sensitivity, Specificity are both common evaluation metrics employed in medical literature and are important in considering the real-life implications of true positives and false negatives. Results from the trained neural network per question are presented in Table 2.

Social media evaluation To perform this, we evaluated our network on the *Reddit* dataset compiled by the authors (Losada and Crestani, 2016). We considered a score above 3 on the PHQ-2 (latter two questions on the PHQ-4) to be indicative of a user suffering depression. Results are presented in Table 3. Accuracy score was 43%, which was 18% below the majority class baseline.

5   Conclusion and Future Work

Exploring new methods to diagnose and treat mental health disorders has become a priority in many countries. Passive diagnosis has the potential to allow for early treatment and diagnosis to become standard practice in society. In the course of this work, we have developed a method to apply this concept to the PHQ-4 to screen for depression and anxiety.

Our approach is the first publication to explore how multilabel neural networks can predict depression and anxiety. We have developed a Multidimensional classification architecture to model the interconnectivity of the symptoms combined with a hardcoded threshold value to output ordinal scores. For multilabel evaluation, our model scores considerably better than the random baseline. While for multidimensional classification our system outperforms a simpler RFC by 14%. When evaluating on social media data from Losada and Crestani (2016), the models fail to match the majority class baseline.

In almost all questions on the PHQ-4, we demonstrate high sensitivity for predicting the disorder. Specificity is slightly lower in many cases, however, for early-stage diagnostics, this is often an acceptable outcome since it often better to ensure false negatives do not occur.

This demonstrates the non-trivial nature of training on one domain of data and evaluating on another. Two out of three of our feature sets, psychometric similarity and text representation employed the pre-trained USE model, which was also trained on non-social media style data. Future work will need to explore the ability to create models that are less semantically domain specific and better able to generalize across writing styles. The concept of transfer learning has seen success in this area (Glorot et al., 2011).

Our approach is incomparable to the proceedings in the *eRisk* workshop who focus on the temporal aspect of the prediction. Data is released in chunks over time and accuracy is penalized as the length of time from the beginning increases.

In final conclusion, our work has demonstrated that neural networks offer a potential new route for the area of passive diagnosis and prediction of depression and anxiety. Future work is required to ensure the generalizability of the approach, however.

Acknowledgments

This publication has emanated from research conducted with the financial support of Science Foundation Ireland (SFI) under Grant Number SFI/12/RC/2289 (Insight).

References

. American Psychiatric Association. 2013. *Diagnostic and Statistical Manual of Mental Disorders, 5th Edition: DSM-5*. American Psychiatric Pub.

Lisa J Barney, Kathleen M Griffiths, Anthony F Jorm, and Helen Christensen. 2006. Stigma about depression and its impact on help-seeking intentions. *Australian & New Zealand Journal of Psychiatry*, 40(1):51–54.

Aaron T. Beck. 1991. Cognitive therapy: A 30-year retrospective. *American Psychologist*, 46(4):368–375.

Daniel Cer, Yinfei Yang, Sheng-yi Kong, Nan Hua, Nicole Limtiaco, Rhomni St John, Noah Constant, Mario Guajardo-Cespedes, Steve Yuan, Chris Tar, et al. 2018. Universal sentence encoder. *arXiv preprint arXiv:1803.11175*.

David A. Clark, Robert A. Steer, and Aaron T. Beck. 1994. Common and Specific Dimensions of Self-Reported Anxiety and Depression: Implications for the Cognitive and Tripartite Models. *Journal of Abnormal Psychology*, 103(4):645–654.

Munmun De Choudhury, Scott Counts, and Eric Horvitz. 2013. Social media as a measurement tool of depression in populations. *Proceedings of the 5th Annual ACM Web Science Conference on - WebSci '13*, pages 47–56.

Munmun De Choudhury and Sushovan De. 2014. Mental Health Discourse on reddit: Self-Disclosure, Social Support, and Anonymity. *Proceedings of the Eight International AAAI Conference on Weblogs and Social Media*, pages 71–80.

Fionn Delahunty, Ian D. Wood, and Mihael Arcan. 2018. First insights on a passive major depressive disorder prediction system with incorporated conversational chatbot. In *CEUR Workshop Proceedings*, volume 2259, pages 327–338.

Klaus P. Ebmeier, Claire Donaghey, and J. Douglas Steele. 2006. Recent developments and current controversies in depression. *Lancet*, 367(9505):153–167.

Frauke Friedrichs and Christian Igel. 2005. Evolutionary tuning of multiple svm parameters. *Neurocomputing*, 64:107–117.

Ouadie Gharroudi, Haytham Elghazel, and Alex Aussem. 2014. A comparison of multi-label feature selection methods using the random forest paradigm. In *Canadian conference on artificial intelligence*, pages 95–106. Springer.

Xavier Glorot, Antoine Bordes, and Yoshua Bengio. 2011. Domain adaptation for large-scale sentiment classification: A deep learning approach. In *Proceedings of the 28th international conference on machine learning (ICML-11)*, pages 513–520.

Jonathan Gratch, Ron Artstein, Gale M Lucas, Giota Stratou, Stefan Scherer, Angela Nazarian, Rachel Wood, Jill Boberg, David DeVault, Stacy Marsella, et al. 2014. The distress analysis interview corpus of human and computer interviews. In *LREC*, pages 3123–3128. Citeseer.

Yen-Hao Huang, Lin-Hung Wei, and Yi-Shin Chen. 2017. Detection of the Prodromal Phase of Bipolar Disorder from Psychological and Phonological Aspects in Social Media.

Robert Johansson, Per Carlbring, Å sa Heedman, Björn Paxling, and Gerhard Andersson. 2013. Depression, anxiety and their comorbidity in the swedish general population: point prevalence and the effect on health-related quality of life. *PeerJ*, 1:e98.

Kurt Kroenke, Robert L Spitzer, Janet BW Williams, and Bernd Löwe. 2010. The patient health questionnaire somatic, anxiety, and depressive symptom scales: a systematic review. *General hospital psychiatry*, 32(4):345–359.

Simon Kung, Renato D Alarcon, Mark D Williams, Kathleen A Poppe, Mary Jo Moore, and Mark A Frye. 2013. Comparing the beck depression inventory-ii (bdi-ii) and patient health questionnaire (phq-9) depression measures in an integrated mood disorders practice. *Journal of affective disorders*, 145(3):341–343.

Klara Latalova, Dana Kamaradova, and Jan Prasko. 2014. Perspectives on perceived stigma and self-stigma in adult male patients with depression. *Neuropsychiatric disease and treatment*, 10:1399.

Yuhua Li, Zuhair A Bandar, and David McLean. 2003. An approach for measuring semantic similarity between words using multiple information sources. *IEEE Transactions on knowledge and data engineering*, 15(4):871–882.

David E Losada and Fabio Crestani. 2016. A Test Collection for Research on Depression and Language Use CLEF 2016, Évora (Portugal). *Experimental IR Meets Multilinguality, Multimodality, and Interaction*, pages 28–29.

Bernd Löwe, Inka Wahl, Matthias Rose, Carsten Spitzer, Heide Glaesmer, Katja Wingenfeld, Antonius Schneider, and Elmar Brähler. 2010. A 4-item measure of depression and anxiety: Validation and standardization of the Patient Health Questionnaire-4 (PHQ-4) in the general population. *Journal of Affective Disorders*, 122(1-2):86–95.

Eneldo Loza Mencia and Johannes Fürnkranz. 2008. Pairwise learning of multilabel classifications with perceptrons. In *IEEE International Joint Conference on Neural Networks*.

Rada Mihalcea, Courtney Corley, Carlo Strapparava, et al. 2006. Corpus-based and knowledge-based measures of text semantic similarity. In *AAAI*, volume 6, pages 775–780.

Jinseok Nam, Jungi Kim, Eneldo Loza Mencía, Iryna Gurevych, and Johannes Fürnkranz. 2014. Large-scale multi-label text classification - Revisiting neural networks. *Lecture Notes in Computer Science (including subseries Lecture Notes in Artificial Intelligence and Lecture Notes in Bioinformatics)*, 8725 LNAI(PART 2):437–452.

James W Pennebaker, Martha E Francis, and Roger J Booth. 2001. Linguistic inquiry and word count: Liwc 2001. *Mahway: Lawrence Erlbaum Associates*, 71(2001):2001.

James W Pennebaker, Matthias R Mehl, and Kate G Niederhoffer. 2003. Psychological aspects of natural language use: Our words, our selves. *Annual review of psychology*, 54(1):547–577.

Trang Pham, Truyen Tran, Dinh Phung, and Svetha Venkatesh. 2016. Deepcare: A deep dynamic memory model for predictive medicine. In *Pacific-Asia Conference on Knowledge Discovery and Data Mining*, pages 30–41. Springer.

Helmut Prendinger and Mitsuru Ishizuka. 2005. The empathic companion: A character-based interface that addresses users' affective states. *Applied Artificial Intelligence*, 19(3-4):267–285.

Faneva Ramiandrisoa and Farah Benamara. 2018. IRIT at e-Risk 2018. In *E-Risk workshop*, pages 367–377.

Jesse Read, Concha Bielza, and Pedro Larrañaga. 2014. Multi-dimensional classification with super-classes. *IEEE Transactions on knowledge and data engineering*, 26(7):1720–1733.

Stephanie S. Rude, Eva Maria Gortner, and James W. Pennebaker. 2004. Language use of depressed and depression-vulnerable college students. *Cognition and Emotion*, 18(8):1121–1133.

Jürgen Schmidhuber. 2015. Deep learning in neural networks: An overview. *Neural networks*, 61:85–117.

Eric J Topol. 2019. High-performance medicine: the convergence of human and artificial intelligence. *Nature medicine*, 25(1):44–56.

Marcel Trotzek, Sven Koitka, and Christoph M. Friedrich. 2018. Utilizing Neural Networks and Linguistic Metadata for Early Detection of Depression Indications in Text Sequences.

. World Health Organization. 2018. Depression Fact Sheet. Technical report.

Min-Ling Zhang and Zhi-Hua Zhou. 2014. A review on multi-label learning algorithms. *IEEE transactions on knowledge and data engineering*, 26(8):1819–1837.

A Appendices

A.1 Hyperparameters

The full pool of possible hyperparameter fed into the genetic algorithm is as follows; possible neurons (64, 128, 256, 768, 2014), possible layers (1, 2, 3, 4, 5, 6, 7), possible activation functions (relu, elu, tanh, sigmoid, hard sigmoid, softplus, linear), possible optimizers (rmsprop, adam, sgd, adagrad, adadelta, adamax, nadam)

HITSZ-ICRC: A Report for SMM4H Shared Task 2019-Automatic Classification and Extraction of Adverse Drug Reactions in Tweets

Shuai Chen[1], Yuanhang Huang[1], Xiaowei Huang[1], Haoming Qin[1], Jun Yan[2], Buzhou Tang[1*]

[1]Department of Computer Science, Harbin Institute of Technology, Shenzhen, China
[2]Yidu Cloud (Beijing) Technology Co., Ltd, Beijing, China
{chenshuai726, hyhang7, kitaharatomoyo, tangbuzhou}@gmail.com
qinhaoming1874@foxmail.com, junyan@yiducloud.cn
*corresponding author

Abstract

This is the system description of the Harbin Institute of Technology Shenzhen (HITSZ) team for the first and second subtasks of the fourth Social Media Mining for Health Applications (SMM4H) shared task in 2019. The two subtasks are automatic classification and extraction of adverse effect mentions in tweets. The systems for the two subtasks are based on bidirectional encoder representations from transformers (BERT), and achieves promising results. Among the systems we developed for subtask1, the best F1-score was 0.6457, for subtask2, the best relaxed F1-score and the best strict F1-score were 0.614 and 0.407 respectively. Our system ranks first among all systems on subtask1.

1 Introduction

Adverse drug reaction (ADR), namely adverse drug effect, is one of the leading causes of post-therapeutic deaths (Saha, Naskar, Dasgupta, & Dey, 2018). Nowadays, more and more people share information in social platform, including health information such as drugs and their ADRs. Twitter, as one of the most popular social platforms, has attracted a great deal of attention from researchers in the medical domain. Some methods, such as HTR_MSA (Wu et al., 2018) and Neural DrugNet (Nikhil & Mundra, 2018), have been proposed to detect tweets mentioning ADRs and medicine intake. In order to facilitate the use of social media for health monitoring and surveillance, the health language processing lab at University of Pennsylvania organized Social Media Mining for Health Applications (SMM4H) shared task four times. In 2019, the fourth SMM4H shared task was comprised of four subtasks: (1) Automatic classifications of adverse effect mentions in tweets,

(2) Extraction of Adverse Effect mentions, (3) Normalization of adverse drug reaction mentions (ADR), and (4) Generalizable identification of personal health experience mentions (Weissenbacher et al., 2019).

We participated in subtask 1 and subtask2, and developed two systems based on bidirectional encoder representations from transformers (BERT) (Devlin, Chang, Lee, & Toutanova, 2018) for the two subtasks respectively. The system for subtask 1 achieved the best F1-score of 0.6457, ranking first. Among the systems we developed for subtask2, the best relaxed F1-score and the best strict F1-score were 0.614 and 0.407 respectively.

2 Task and Data Description

2.1 Task 1: Automatic Classifications of Adverse Effect Mentions in Tweets

Task 1 was formulated as follows: given a tweet, determine whether it mentions drug adverse effect mentions, denoted by 1 and 0, indicating a tweet mentions drug adverse effects and not, respectively. The organizers provided a train dataset consisting of 25,678 tweets for all participants to develop their system, and a test dataset consisting of 4,575 tweets to evaluate the performance of all systems. Table 1 shows the distribution of 0 and 1 labels over the training and test datasets, where #* denotes the number of tweets labeled with *, and NA denotes that the corresponding number is currently unknown.

Dataset	#1	#0	#all
Training set	2,377	23,301	25,678
Test set	NA	NA	4,575

Table 1: Distribution of labels over the training and test datasets of task1.

Proceedings of the 4th Social Media Mining for Health Applications (#SMM4H) Workshop & Shared Task, pages 47–51
Florence, Italy, August 2, 2019. ©2019 Association for Computational Linguistics

2.2 Task 2: Extraction of Adverse Effect Mentions

Task 2 as a follow-step of Task 1 was formulated as follows: given a tweet, identify the text span of adverse effect mentions. The challenge of task 2 is to distinguish adverse effect mentions from similar non-ADR expressions. A training set of 3,225 tweets annotated with 1830 adverse effect mentions was provided for system development, and a test set of 1,573 tweets was provided for system evaluation. The statistics of the training and test datasets are listed in Table 2.

Dataset	#tweets	#ADRs
Training set	3,225	1,830
Test set	1,573	NA

Table 2: Statistics of the training and test datasets of task 2

3 Methods

Our systems for both task 1 and task 2 were based on BERT, an unsupervised language representation method to obtain deep bidirectional representations of sentences by jointly conditioning on both left and right context in all layers from free text. Below we described in detail the methods for the two tasks: task 1 and task 2, respectively.

3.1 Task 1: BERT and BERT+Knowledge Base

In this task, we designed two methods, BERT and BERT +Knowledge Base. The model architecture is shown in Fig. 1.

BERT: Like what BERT did, we took the final hidden state of the first input token [CLS] as the representation of a tweet. Then we applied a softmax layer over the output to classify a tweet. We denote the representation vector as H, then the predicted label $\hat{y}$ is computed as:

$$\hat{y} = softmax(WH + b) \qquad (1)$$

where W, b is the parameters of the fully connected layer.

BERT+Knowledge Base: Inspired by Li et al. (2018), we tried to combine the BERT output with features from knowledge bases to improve the performance of systems. We firstly extracted drugs which appear in the SIDER 4.1 (a side effect resource which contains information on marketed medicines and their recorded adverse drug reactions) from the train dataset, and obtained a drug lexicon of 538 drugs. Then we extracted corresponding adverse effects in SIDER according to the drug lexicon, and obtained 4,411 <drug, ADR> pairs. For each tweet, according to the presence of <drug ADR> pairs, we could build a binary feature. We incorporated the binary feature into representation vectors of a tweet. The final representation of a tweet is a concatenation of its BERT output and lexicon feature. Then we used a fully connected layer to fuse information from different feature spaces, and applied a softmax layer on it to classify tweets. We denote the output of BERT as H_1, the lexicon feature as H_2, then the predicted label $\hat{y}$ of a tweet is computed as :

$$\hat{y} = softmax(W[H_1, H_2] + b) \qquad (2)$$

where W, b is the parameters of the fully connected layer. The loss function for two models training is crossentropy:

$$L = -\sum_{i=1}^{N} \sum_{j=1}^{C} y_{ij} \cdot log(\hat{y}_{ij}) \qquad (3)$$

Where y_{ij} and $\hat{y}_{ij}$ are gold label and predicted label for the i_{th} sample in the j_{th} label category. N is the number of samples in a batch, C is the number of label categories.

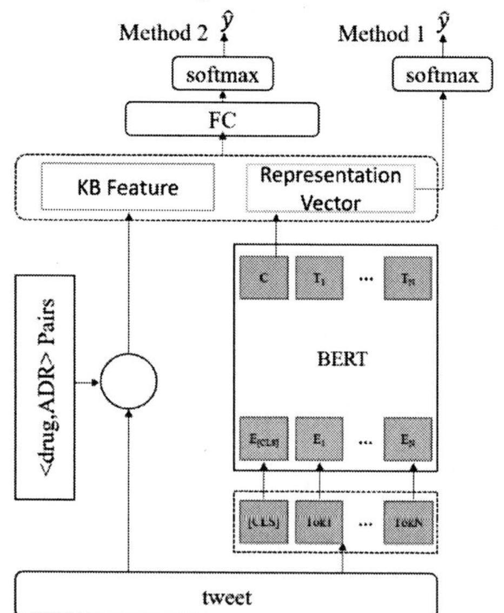

Figure 1: The model architecture in Task 1

3.2 Task 2: BERT and BERT+CRF

In task2, we still took BERT as the basic architecture, and designed two methods. The model architecture is shown in Fig. 2.

BERT: This method is very similar to the first method in Task 1. The difference is that we feed the final hidden representation for to each token into a classification layer over the NER tags set, because we need to obtain predicted tag of each input token.

BERT +CRF: This method is a follow step of the first method. For BERT method, the predictions are not conditioned on the surrounding predictions. A CRF layer has a state transition matrix as parameters (Huang, Xu, & Yu, 2015). With such a layer, the system can efficiently use past and future tags to predict the current tag. Therefore, we applied a CRF layer on the classification layer. We denote the output sequence after softmax layer as $H = [h_1, h_2, \dots h_n]$, then the predicted tag sequence $Z = [z_1, z_2, \dots z_n]$ is as follows:

$$Z = \underset{y}{\arg\max} \frac{exp(score(H,y))}{\sum_{y'} exp(score(H,y'))} \quad (4)$$

where $score(H,y) = \sum_{t=1}^{n} E_{t,y_t} + \sum_{t=0}^{n-1} T_{y_t y_{t+1}}$, $E_{t,y_t} = w_{y_t}^{\mathrm{T}} h_t$ is the score of predicting tag y_t at the t_{th} time, and $T_{y_t y_{t+1}}$ is the score of transitioning from y_t to y_{t+1}.

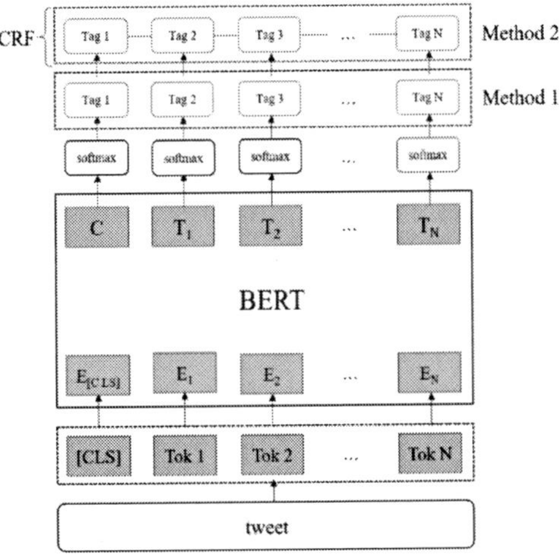

Figure 2: The model architecture in Task 2

3.3 Experiments

For task1, we compared BERT and BERT+knowledge base with two classic deep learning methods, TextCNN (Kim, 2014) and LSTM (Hochreiter & Schmidhuber, 1997), and also investigated the effect of different BERT models, including the BERT model (Devlin et al., 2018) publicly released by (https://github.com/google-research/bert) (denoted by BERT_noRetrained) and the BERT model retrained on a large-scale tweet unlabeled corpus based on the previous BERT model (denoted by BERT_Retrained). The unlabeled corpus consisted of 1,500,000 tweets crawled from Twitter according to 150 drug names collected from the training set. For task2, we only used the retrained BERT model.

In our experiments, we set batch size to 32, learning rate to 5e-5 when training all models. The epoch number was set to 8 for BERT retraining, and 20 for other models. The dimension of word embeddings used in TextCNN and LSTM was set to 200. We split out about 10% from the training set as a validation set for parameter optimization. The performance of all methods for the two tasks were measured by precision, recall and F1-score, which can be calculated by the official tools provided by the organizers. For task2, there were two criteria for system performance evaluation: relaxed and strict.

4   Results

Table 3 and Table 4 show the performance of our systems for task 1 and task 2 on the test set, respectively.

For task 1, among the systems we developed, "BERT_Retrained" achieved the best F1-score of 0.6457 and recall of 0.6885 on the test set, "BERT_Retrained+Knowledge Base" achieved the best precision of 0.6916 on the test set. Compared with TextCNN and LSTM on the validation set, methods based on BERT showed much better performance. As officially reported, "BERT_Retrained" ranked first among all systems.

For task 2, among the systems we developed, "BERT_Retrained+CRF" achieved the best relaxed F1-score of 0.614 and the best strict F1-score of 0.407, outperforming "BERT_Retrained" by 0.024 in relaxed F1-score and 0.060 in strict F1-score.

5   Discussion

For Task 1, the distribution of 0 and 1 is highly imbalanced, 90% of samples are negative, 10% of samples are positive. When we used CNN and LSTM, if we did not deal with the data imbalance problem, the performance of them was quite poor, most tweets were classified to 0. In order to balance the number of positive and negative samples, we randomly divided into the negative

System	Validation			Test		
	F1	P	R	F1	P	R
TextCNN	0.491	0.464	0.522		\	
LSTM	0.483	0.516	0.453		\	
BERT_noRretrained	0.618	0.646	0.593		\	
BERT_Retrained	**0.665**	0.611	**0.728**	**0.6457**	0.6079	**0.6885**
BERT_Retrained+Knowledge Base	0.642	**0.720**	0.579	0.6289	**0.6916**	0.5767
Average of participants' systems	\	\	\	0.5019	0.5351	0.5054

Table 3: Results on validation and test data for Task 1

System	Relaxed			Strict		
	F1	P	R	F1	P	R
BERT_Retrained+CRF	**0.614**	0.538	0.716	**0.407**	0.357	0.474
BERT_Retrained	0.59	0.529	0.666	0.347	0.311	0.392
Average of participants' systems	0.5383	0.5129	0.6174	0.3169	0.3026	0.3581

Table 4: Results on test data for Task 2

samples into five equal parts, and combined each part with the positive samples to form a new training dataset. After this operation, we obtained five new balanced training datasets. Then we trained five models on them, and ensembled the five models. The ensembled model brought an increase of about 8% in F1-score. However, when applying this operation to BERT and "BERT+Retrained", we obtained little increase on F1-score.

By analyzing results of "BERT_Retrained", we found that the main errors are:

- ADR mentions cannot be compeletely distinguished from the reason mentions of taking drugs. For example, in "oxycodone just took my headache away so fast", "headache" is the reason of taking oxycodone, not an adverse effect mention of oxycodone. The tweet was wrongly classified to 1.

- Implicit adverse effect mentions are difficult to identified. For example, "pristiq and im livin in a cold world" and "uhh my gabapentin does went up today and I don't even know what planet i'm on. i hope i adjust to this quickly ... #endometriosis".

For task 2, because the CRF layer takes full advantages of relations between neighbor labels, "BERT_Retrained+CRF" could avoid some terrible tag sequences such as "I-B-B-O-O". The main errors appearing in task 2 are the same as task 1.

For further improvement, a possible direction is dealing with task 1 and task 2 at the same time using joint learning methods.

6 Conclusion

In this paper, we developed systems for task 1 and task 2 of the SMM4H shared task in 2019. Our systems were based on BERT and achieved promising results, especially ranking first on task 1.

References

Devlin, J., Chang, M.-W., Lee, K., & Toutanova, K. (2018). Bert: Pre-training of deep bidirectional transformers for language understanding. *ArXiv Preprint ArXiv:1810.04805*.

Hochreiter, S., & Schmidhuber, J. (1997). Long short-term memory. *Neural Computation, 9*(8), 1735–1780.

Huang, Z., Xu, W., & Yu, K. (2015). Bidirectional LSTM-CRF Models for Sequence Tagging. *ArXiv Preprint ArXiv:1508.01991*.

Kim, Y. (2014). Convolutional neural networks for sentence classification. *ArXiv Preprint ArXiv:1408.5882*.

Li, H., Yang, M., Chen, Q., Tang, B., Wang, X., & Yan, J. (2018). Chemical-induced disease extraction via recurrent piecewise convolutional neural networks. *BMC Medical Informatics and Decision Making, 18*(2), 60.

Nikhil, N., & Mundra, S. (2018, October). Neural DrugNet. *Proceedings of the 2018 EMNLP Workshop SMM4H: The 3rd Social Media Mining for Health Applications Workshop and Shared Task.*

Saha, R., Naskar, A., Dasgupta, T., & Dey, L. (2018). Leveraging Web Based Evidence Gathering for Drug Information Identification from Tweets. *Proceedings of* the *2018 EMNLP Workshop SMM4H: The 3rd Social Media Mining for Health Applications Workshop & Shared Task.*

Weissenbacher, D., Sarker, A., Magge, A., Daughton, A., O'Connor, K., Paul, M., & Graciela Gonzalez-Hernandez. (2019). Overview *of* the Fourth Social Media Mining for Health (SMM4H) Shared Task at ACL 2019. *Proceedings of the 2019 ACL Workshop SMM4H: The 4th Social Media Mining for Health Applications Workshop & Shared Task.*

Wu, C., Wu, F., Liu, J., *Wu,* S., Huang, Y., & Xie, X. (2018, October). Detecting Tweets Mentioning Drug Name and Adverse Drug Reaction with Hierarchical Tweet Representation and Multi-Head Self-Attention. *Proceedings of the 2018 EMNLP Workshop SMM4H: The 3rd Social Media Mining for Health Applications Workshop and Shared Task.*

KFU NLP Team at SMM4H 2019 Tasks: Want to Extract Adverse Drugs Reactions from Tweets? BERT to The Rescue

Zulfat Miftahutdinov and **Ilseyar Alimova**
Kazan Federal University,
Kazan, Russia
zulfatmi@gmail.com
ISAlimova@kpfu.ru

Elena Tutubalina
Kazan Federal University,
Kazan, Russia
Samsung-PDMI Joint AI Center,
PDMI RAS, St. Petersburg, Russia
elvtutubalina@kpfu.ru

Abstract

This paper describes a system developed for the Social Media Mining for Health (SMM4H) 2019 shared tasks. Specifically, we participated in three tasks. The goals of the first two tasks are to classify whether a tweet contains mentions of adverse drug reactions (ADR) and extract these mentions, respectively. The objective of the third task is to build an end-to-end solution: first, detect ADR mentions and then map these entities to concepts in a controlled vocabulary. We investigate the use of a language representation model BERT trained to obtain semantic representations of social media texts. Our experiments on a dataset of user reviews showed that BERT is superior to state-of-the-art models based on recurrent neural networks. The BERT-based system for Task 1 obtained an F1 of 57.38%, with improvements up to +7.19% F1 over a score averaged across all 43 submissions. The ensemble of neural networks with a voting scheme for named entity recognition ranked first among 9 teams at the SMM4H 2019 Task 2 and obtained a relaxed F1 of 65.8%. The end-to-end model based on BERT for ADR normalization ranked first at the SMM4H 2019 Task 3 and obtained a relaxed F1 of 43.2%.

1 Introduction

Short-text communication forms, such as Twitter microblogging, present a wide variety of facts and opinions on numerous topics, and this treasure trove of information is currently severely underexplored. Here we focus on the problem of discovering adverse drug reaction (ADR) concepts in Twitter messages as part of the Social Media Mining for Health (SMM4H) 2019 shared tasks.

This work is based on the participation of our team, named *KFU NLP*, in the first three tasks. Organizers of SMM4H 2019 Tasks 1-3 (Weissenbacher et al., 2019) provided participants with datasets of English tweets annotated at the message level with binary annotation indicating the presence or absence of ADRs, text spans of reported ADRs, and their corresponding medical codes from the Medical Dictionary for Regulatory Activities (MedDRA). The goal of Task 1 is to classify the tweets according to the presence of ADRs. For the second task, named entity recognition (NER) aims to detect the mentions of ADRs. The third and final task is designed as an end-to-end problem, intended to perform full evaluation of a system operating in real conditions: given a set of raw tweets, the system has to find the tweets that are mentioning ADRs, find the spans of the ADRs, and normalize them with respect to a given knowledge base (KB). These tasks are especially challenging due to specific characteristics of user-generated texts from social networks which are noisy, containing misspelled words, abbreviations, emojis, etc.

Motivated by the recent success of deep architectures in general and language representation networks in particular, we explore an application of Bidirectional Encoder Representations from Transformers (BERT) (Devlin et al., 2018) and its extension for biomedical domain BioBERT (Lee et al., 2019) to the SMM4H 2019 tasks. For both ADR extraction and medical concept normalization, we conclude that BERT outperforms previous state-of-the-art baselines based on recurrent neural architectures (RNNs), including bidirectional Long Short-Term Memory (LSTM) (Hochreiter and Schmidhuber, 1997), and Gated Recurrent Units (Cho et al., 2014) paired with *word2vec* word embeddings.

The paper is organized as follows. In Section 2, we present the task description, machine learning baselines, and classification experiments for Task 1. We describe our models for end-to-end extraction of ADR concepts in Sections 3 and 4. Finally,

Proceedings of the 4th Social Media Mining for Health Applications (#SMM4H) Workshop & Shared Task, pages 52–57
Florence, Italy, August 2, 2019. ©2019 Association for Computational Linguistics

we discuss future directions in Section 5.

2   Task 1: Classification

The goal of this sub-task is to identify the tweets with ADR mentions. This is a necessary filtering step to remove noise, since most of the health-related chatter in the domain does not contain relevant information.

2.1   Dataset

The training set consists of 25,678 tweets with 2,377 labeled as positive examples with ADRs; this statistic shows that the corpus has huge class imbalance. Tweet text lengths vary from 1 to 53 words, the average length is 20 words. The test dataset includes 4,576 tweets. Minimum tweet length is also 1 and the maximum consists of 186 words, which is much longer than in the training set. However, the average amount of words in tweets is on par with the training set and equals 23 words.

2.2   Method

Previous studies have shown the effectiveness of classical machine learning approaches (Ofoghi et al., 2016; Jonnagaddala et al., 2016; Kiritchenko et al., 2018; Alimova and Tutubalina, 2017). We applied the SVM-based model with a set of features as a baseline method. For SVM features, we utilized the bag-of-words representation, drug name, and ADRs from a Diego Lab ADR lexicon (Sarker et al., 2015). The list of drug names was obtained from the Food and Drug Administration (FDA). We've also explored the potential of *sent2vec* tool for tweets representation (Pagliardini et al., 2018). The Twitter unigram pre-trained model was applied for obtaining vectors[1].

Our main solution is a classifier based on the BERT architecture. For the BERT-based model, the tweet's representation was obtained with the *Transformer* architecture (Vaswani et al., 2017), and then logistic regression was used as a classifier. We used the implementation from the model's official repository[2].

2.3   Experiments

For the SVM-based classifier, we set class weights to 0.3 and 0.7 for non-ADR and ADR classes re-

Run name	F1	P	R
KFU NLP, BERT	57.38	69.14	49.04
KFU NLP, SVM	51.64	56.2	47.76
Average scores	50.19	53.51	50.54

Table 1: Text classification results on the Task 1 test set.

spectively and applied a linear kernel. The BERT-based model was trained on 20 epochs with learning rate equal to $5*10^{-5}$, maximum sequence size 128, and batch size 32.

The official evaluation metrics are precision (P), recall (R), and F1-measure (F1) computed for the positive class. During preprocessing, we removed all URLs, user mentions, and symbols of re-tweets using the *tweet-preprocessor* package[3]. We conducted a set of experiments on the training set with 5-fold cross-validation. Results of these experiments shows that utilizing *sent2vec* as tweet representations did not improve classification quality. Results on the test set are presented in Table 1. Our baseline SVM classifier (run-2) obtained the F1 score of 51.64%, which is on par with average results. The BERT-based classifier (run-1) achieved the F1 score of 57.38 and outperformed by 7.19% the F1 score averaged across 43 submissions.

3   Task 2: Extraction of Adverse Effect Mentions

Following state-of-the-art research (Miftahutdinov et al., 2017; Tutubalina and Nikolenko, 2017; Lee et al., 2019), we view the second task from the perspective of a sequence labeling problem. Sequence labeling refers to the task of learning to predict a label for each token in a sequence of tokens. State-of-the-art methods employ neural architectures based on bidirectional LSTMs and conditional random fields (CRF) (Lample et al., 2016; Tutubalina and Nikolenko, 2017; Giorgi and Bader, 2019). Recent advancements in language representation models such as BERT have opened up new directions of research in sequence labeling.

3.1   Dataset

The data for the second sub-task includes 2,367 tweets that are fully annotated for ADR mentions and Indications. This set contains a subset of (i) 1,212 tweets from Task 1 tagged as 'hasADR' and

---

[1] https://github.com/epfml/sent2vec
[2] https://github.com/google-research/bert

[3] https://pypi.org/project/tweet-preprocessor/

(ii) 1,155 tweets marked as 'noADR' (1,828 ADR mentions in total).

3.2 Method

Sequence labeling methods view a message as a sequence of tokens labeled using the BIO tagging scheme: B indicates the beginning of the entity mention, I is used for tokens inside the entity mention, and O indicates tokens outside any entities. To solve the sequence labeling task, we utilize and empirically compare several models: (i) bidirectional LSTM-CRF; (ii) BERT; (iii) BERT for Biomedical Text Mining named BioBERT. We have also utilized a CRF tagger on top of BioBERT. A technical explanation of these neural models is omitted due to space constraints; we refer to the studies listed above.

We have also combined deep neural network representations with additional dictionary-based features. Dictionary-based features are calculated for each token in a text as follows: first, all the occurrences of predefined vocabulary entries were found in the text, then the first token of the matched part tagged was with B-tag, the last with I-tag, and all other tokens in the text with O-tag. The dictionary-based features are concatenated with the representation learned by the neural network that captures extensional semantic information of an entity mention. We adopted the dictionaries from our previous work (Miftahutdinov et al., 2017).

3.3 Experiments

For the NER sub-task each network was trained for 25 epochs with batch size set to 32. We used the Adam algorithm as the optimizer with initial learning rate $5 * 10^{-5}$. We used the publicly available implementation of BioBERT-CRF[4]. Training all 10 networks took 2-3 hours on eight NVIDIA Tesla P40 GPUs. Additionally, we have used the CADEC corpus along with the corpus provided by the organizers.

Since the boundaries of an entity mention in social media texts are hard to define, two types of evaluation were used: *strict* and *relaxed*. Precision, recall, and F-measure are used for performance evaluation.

In order to select the best neural models, we evaluated our models on the CADEC corpus using 5-fold cross-validation at the develop-

---

Run name	F1	P	R
Relaxed Evaluation			
KFU NLP Team	65.8	55.4	81.0
Average scores	53.83	51.29	61.74
Strict Evaluation			
KFU NLP Team	46.4	38.9	57.6
Average scores	31.69	30.26	35.81

Table 2: The NER results on the Task 2 test set.

ment stage. BERT showed 5-7% improvement in the strict evaluation over LSTM-CRF, while BioBERT showed slightly better performance over BERT. BioBERT with CRF stayed roughly on par with the model without CRF.

During BioBERT evaluation, we encountered unstable results on development sets. Therefore, for the final submission we combined the results of ten BioBERT-CRF with the same settings using a simple voting scheme with the intent of increasing the robustness of the final system. Table 2 shows a comparison of the ensemble model to the official average scores computed using the participants' submissions. Our model has obtained the highest relaxed F1 score of 65.8% among 9 teams.

4 Task 3: Medical Concept Normalization

A crucial part of this problem is to translate a text from *social media language* (e.g., "felt sick to my stomach" or "couldn't sleep much") to *formal medical language* (e.g., "nausea" and "insomnia", respectively).

The SMM4H 2019 Task 3 is designed as an end-to-end task. This setup is closer to a real production environment, where the system has free-form text as input and should be able to produce a set of extracted medical concepts. This end-to-end setup is more challenging due to the sequential two-stage pipeline: the system has to (i) first detect ADR mentions and then (ii) map extracted ADRs to knowledge base entries. For the first step, we use the NER model described in Section 3. The system used for concept normalization is based on our previous works (Tutubalina et al., 2018; Miftahutdinov and Tutubalina, 2019) and presented below.

4.1 Dataset

ADR mentions from the SMM4H 2019 dataset are mapped to Preferred Terms (PTs) of the Medical

---

⁴https://github.com/dmis-lab/biobert

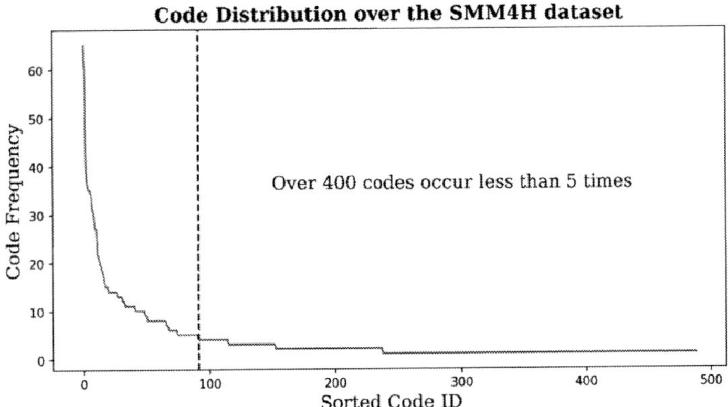

Figure 1: The code frequency distribution of MedDRA codes in the training set (Task 3).

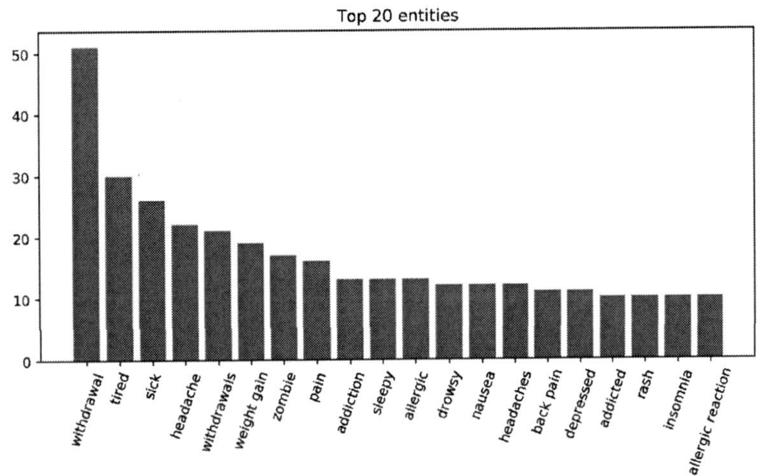

Figure 2: Top 20 entities in the training set (Task 3).

Dictionary for Regulatory Activities (MedDRA). The training SMM4H 2019 set consists of 1,828 phrases mapped to 489 MedDRA codes. The average number of ADR mentions mapped to a given concept is 3.74. The minimum and maximum numbers of queries mapped to a given concept are 1 and 65, respectively. Figure 1 shows a plot of the code frequency distribution of MedDRA concepts presented in the training set. Additionally, we present statistics on the top 20 entity mentions from the training set in Figure 2.

4.2 Method

Following state-of-the-art research (Tutubalina et al., 2018; Sarker et al., 2018; Miftahutdinov and Tutubalina, 2019), we view concept normalization as a classification task. Following (Miftahutdi-nov and Tutubalina, 2019), we convert each ADR mention into a vector representation using BERT or RNN. Next, we employ the standard softmax activation for the output layer. The softmax layer over all possible medical codes from the training set yields a probability for the sequence.

In order to train the classification model, we utilized training sets from five different sources: SMM4H 2019 dataset, SMM4H 2017 dataset (Sarker et al., 2018), CADEC dataset (Karimi et al., 2015), PsyTAR dataset (Zolnoori et al., 2019), and TwADR-L (Limsopatham and Collier, 2016). SMM4H datasets and CADEC were manually mapped to MedDRA codes. PsyTAR and TwADR-L were mapped to the MedDRA coding system using the UMLS metathesaurus (version 2017AA).

Run name	F1	P	R
Relaxed Evaluation			
KFU NLP Team	43.2	36.2	53.5
Average scores	29.72	29.06	31.15
Strict Evaluation			
KFU NLP Team	34.4	28.8	42.7
Average scores	21.18	20.53	22.41

Table 3: The concept normalization results on the Task 3 test set.

4.3 Experiments

We trained the BERT model for 40 epochs, using batch size 96 and learning rate $5*10^{-5}$. In order to prevent neural networks from overfitting, we used a dropout of 0.2 to control the inputs and the softmax layer. We used the publicly available implementation of BERT[5].

The strict and relaxed evaluations proposed for Task 2 were also adopted for Task 3. As in previous work, we evaluated our models on the CADEC corpus at the development stage using 5-fold cross-validation. The BERT model consistently outperformed attention-based bidirectional LSTM and GRU paired with pre-trained word embeddings in this set of experiments, showing a 6-9% improvement. We did not experiment with BioBERT for this task.

For the final submission, we used the two-stage pipeline based on the ensemble of BioBERT-CRF for NER and BERT for normalization. Table 3 shows a comparison of our best model to the official average scores computed using the participants' submissions. The end-to-end model ranked first at SMM4H 2019 Task 3 and obtained a relaxed F1 of 43.2%. The strict recall of the end-to-end system is 15% lower than the recall of the NER system: 42.7 vs 57.6. Results in Tables 2 and 3 indicate that more than 80% of extracted ADR mentions have been correctly mapped to MedDRA concepts.

5 Conclusion

In this work, we have explored an application of Bidirectional Encoder Representations from Transformers (BERT) to the task of text classification, extraction of adverse drug reactions, and concept normalization. We have evaluated BERT and BioBERT empirically against bidirectional LSTM

and GRU. Experiments have shown that BERT outperforms LSTM and GRU on all three tasks, achieving new state-of-the-art results in ADR extraction and normalization.

We foresee three directions for future work. One potential direction is to investigate neural architectures including BERT and RNNs in the end-to-end setup on other existing corpora. Another future direction is to explore how to effectively use of contextual information to map entity mentions to medical concepts. Additionally, the effect of data imbalance can be investigated for BERT-based models.

Acknowledgments

We thank Sergey Nikolenko for helpful discussions. This research was supported by the Russian Science Foundation grant no. 18-11-00284.

References

Ilseyar Alimova and Elena Tutubalina. 2017. Automated detection of adverse drug reactions from social media posts with machine learning. In *International Conference on Analysis of Images, Social Networks and Texts*, pages 3–15. Springer.

Kyunghyun Cho, Bart van Merrienboer, Çaglar Gülçehre, Fethi Bougares, Holger Schwenk, and Yoshua Bengio. 2014. Learning phrase representations using RNN encoder-decoder for statistical machine translation. *CoRR*, abs/1406.1078.

Jacob Devlin, Ming-Wei Chang, Kenton Lee, and Kristina Toutanova. 2018. Bert: Pre-training of deep bidirectional transformers for language understanding. *arXiv preprint arXiv:1810.04805*.

John Giorgi and Gary Bader. 2019. Towards reliable named entity recognition in the biomedical domain. *BioRxiv*, page 526244.

S. Hochreiter and J. Schmidhuber. 1997. Long Short-Term Memory. *Neural Computation*, 9(8):1735–1780. Based on TR FKI-207-95, TUM (1995)

Jitendra Jonnagaddala, Toni Rose Jue, and Hong-Jie Dai. 2016. Binary classification of twitter posts for adverse drug reactions. In *Proceedings of the Social Media Mining Shared Task Workshop at the Pacific Symposium on Biocomputing, Big Island, HI, USA*, pages 4–8.

Sarvnaz Karimi, Alejandro Metke-Jimenez, Madonna Kemp, and Chen Wang. 2015. Cadec: A corpus of adverse drug event annotations. *Journal of biomedical informatics*, 55:73–81.

Svetlana Kiritchenko, Saif M Mohammad, Jason Morin, and Berry de Bruijn. 2018. Nrc-canada at

---

[5]`https://github.com/huggingface/pytorch-pretrained-BERT`

smm4h shared task: classifying tweets mentioning adverse drug reactions and medication intake. *arXiv preprint arXiv:1805.04558*.

Guillaume Lample, Miguel Ballesteros, Sandeep Subramanian, Kazuya Kawakami, and Chris Dyer. 2016. Neural architectures for named entity recognition. In *Proceedings of NAACL-HLT*, pages 260–270.

Jinhyuk Lee, Wonjin Yoon, Sungdong Kim, Donghyeon Kim, Sunkyu Kim, Chan Ho So, and Jaewoo Kang. 2019. Biobert: pre-trained biomedical language representation model for biomedical text mining. *arXiv preprint arXiv:1901.08746*.

Nut Limsopatham and Nigel Collier. 2016. Normalising Medical Concepts in Social Media Texts by Learning Semantic Representation. In *ACL*.

Z.Sh. Miftahutdinov, E.V. Tutubalina, and A.E. Tropsha. 2017. Identifying disease-related expressions in reviews using conditional random fields. *Komp'juternaja Lingvistika i Intellektual'nye Tehnologii*, 1(16):155–166.

Zulfat Miftahutdinov and Elena Tutubalina. 2019. Deep neural models for medical concept normalization in user-generated texts. In *Proceedings of ACL 2019, Student Research Workshop*, Florence, Italy. Association for Computational Linguistics.

Bahadorreza Ofoghi, Samin Siddiqui, and Karin Verspoor. 2016. Read-biomed-ss: Adverse drug reaction classification of microblogs using emotional and conceptual enrichment. In *Proceedings of the Social Media Mining Shared Task Workshop at the Pacific Symposium on Biocomputing*.

Matteo Pagliardini, Prakhar Gupta, and Martin Jaggi. 2018. Unsupervised Learning of Sentence Embeddings using Compositional n-Gram Features. In *NAACL 2018 - Conference of the North American Chapter of the Association for Computational Linguistics*.

Abeed Sarker, Maksim Belousov, Jasper Friedrichs, Kai Hakala, Svetlana Kiritchenko, Farrokh Mehryary, Sifei Han, Tung Tran, Anthony Rios, Ramakanth Kavuluru, et al. 2018. Data and systems for medication-related text classification and concept normalization from twitter: insights from the social media mining for health (smm4h)-2017 shared task. *Journal of the American Medical Informatics Association*, 25(10):1274–1283.

Abeed Sarker, Rachel Ginn, Azadeh Nikfarjam, Karen OConnor, Karen Smith, Swetha Jayaraman, Tejaswi Upadhaya, and Graciela Gonzalez. 2015. Utilizing social media data for pharmacovigilance: a review. *Journal of biomedical informatics*, 54:202–212.

E. Tutubalina and S. Nikolenko. 2017. Combination of deep recurrent neural networks and conditional random fields for extracting adverse drug reactions from user reviews. *Journal of Healthcare Engineering*, 2017.

Elena Tutubalina, Zulfat Miftahutdinov, Sergey Nikolenko, and Valentin Malykh. 2018. Medical concept normalization in social media posts with recurrent neural networks. *Journal of biomedical informatics*, 84:93–102.

Ashish Vaswani, Noam Shazeer, Niki Parmar, Jakob Uszkoreit, Llion Jones, Aidan N Gomez, Ł ukasz Kaiser, and Illia Polosukhin. 2017. Attention is all you need. In I. Guyon, U. V. Luxburg, S. Bengio, H. Wallach, R. Fergus, S. Vishwanathan, and R. Garnett, editors, *Advances in Neural Information Processing Systems 30*, pages 5998–6008. Curran Associates, Inc.

Davy Weissenbacher, Abeed Sarker, Arjun Magge, Ashlynn Daughton, Karen O'Connor, Michael Paul, and Graciela Gonzalez-Hernandez. 2019. Overview of the fourth social media mining for health (smm4h) shared task at acl 2019. In *Proceedings of the 2019 ACL Workshop SMM4H: The 4th Social Media Mining for Health Applications Workshop & Shared Task*. Association for Computational Linguistics.

Maryam Zolnoori, Kin Wah Fung, Timothy B Patrick, Paul Fontelo, Hadi Kharrazi, Anthony Faiola, Yi Shuan Shirley Wu, Christina E Eldredge, Jake Luo, Mike Conway, et al. 2019. A systematic approach for developing a corpus of patient reported adverse drug events: A case study for ssri and snri medications. *Journal of biomedical informatics*, 90:103091.

Approaching SMM4H with Merged Models and Multi-task Learning

Tilia Ellendorff, Lenz Furrer, Nicola Colic, Noëmi Aepli, Fabio Rinaldi
Institute of Computational Linguistics, University of Zurich
`{name.surname}@uzh.ch`

Abstract

We describe our submissions to the 4th edition of the Social Media Mining for Health Applications (SMM4H) shared task. Our team (UZH) participated in two sub-tasks: *Automatic classifications of adverse effects mentions in tweets* (Task 1) and *Generalizable identification of personal health experience mentions* (Task 4). For our submissions, we exploited ensembles based on a pretrained language representation with a neural transformer architecture (BERT) (Tasks 1 and 4) and a CNN-BiLSTM(-CRF) network within a multi-task learning scenario (Task 1). These systems are placed on top of a carefully crafted pipeline of domain-specific pre-processing steps.

1 Introduction

The Social Media Mining for Health Applications (SMM4H) shared task 2019 (Weissenbacher et al., 2019) focused on classical natural-language-processing (NLP) problems applied to Twitter microposts (tweets). Our team participated in two tasks of binary text classification: tweets are labeled positive if they contain an Adverse Drug Reaction (ADR) in Task 1 or a Personal Health Mention (PHM) in Task 4. Task 1 (automatic classifications of adverse effects mentions in tweets) is a re-run of the ADR task from previous editions of the SMM4H shared task. Task 4 (generalizable identification of personal health experience mentions) was run for the first time. This task consists in deciding if a tweet contains personal health mentions, as opposed to mentions of general awareness of a health issue. Here, the main challenge is to generalize from the health contexts given by the two datasets provided as training data (i.e. flu vaccination and flu infection) to other, possibly very different, health contexts.

2 Data and Pre-processing

The organizers provided all participants with labeled training data which included the text of the tweets (as opposed to the previous years where only tweet ids were provided). Table 1 describes the size of the available datasets.

Data for Task 4 originated from two different flu-related contexts, namely flu infection (Lamb et al., 2013) and flu vaccination (Huang et al., 2017). Each of these two datasets has their own specific scope. Within the infection dataset, positively labeled examples are restricted to reports of own infection (i.e., the author of the tweet is infected) or infection of somebody close to the author, whereas tweets mentioning personal vaccination are labeled as negative. The vaccination dataset labels tweets as positive only if they report that either the author, or a person close to the author, has actually been vaccinated. Tweets about personal infection are labeled as negative within this dataset. Task 4, on the other hand, looks to label all instances as positive which contain a personal health mention (be it infection or vaccination or any other health context) without a specified restricted scope. Therefore, the main challenge of Task 4 is to generalize from the specific health contexts, as provided within the training data, to personal health mentions in general.

For both tasks, we pre-processed all tweets with the following steps:

- Without sentence splitting, the tweets are tokenized using NLTK's Twitter tokenizer.[1]

- User names and numbers are replaced with "@user" and "NUMBER", respectively.

- URLs are truncated to their domain names.

- Hash symbols are stripped from hash tags.

---

[1] `https://www.nltk.org/api/nltk.tokenize.html`

Proceedings of the 4th Social Media Mining for Health Applications (#SMM4H) Workshop & Shared Task, pages 58–61
Florence, Italy, August 2, 2019. ©2019 Association for Computational Linguistics

		# tweets			# unique tweets		
		neg	pos	total	neg	pos	total
Task 1	total	23301	2377	25678	22497	2368	24861
Task 4	*Inf*	472	564	1036	460	545	1005
	Vacc	4815	1900	6715	4680	1885	6515
	total	5287	2464	7751	5140	2430	7570

Table 1: Number of tweets provided for each task as training data. Task 4 includes data from the health context of Vaccination (*Vacc*) and Infection (*Inf*). Unique tweets are counted after pre-processing followed by removal of duplicates.

- Camel-cased expressions like "SideEffects" are split into their component words.

- Artifacts of upstream processing like "&" are fixed.

- Frequent colloquial abbreviations (e.g. "w/" for "with") are resolved.

- Repeated letters ("greaaaaat") are removed. Specifically, runs of three or more equal letters are replaced with a single occurrence, except for "e", where two letters are retained (e.g. "freeeeeze" becomes "freeze"). Letter de-repetition was not applied to the BERT-based systems (described below).

The datasets contain a considerable number of duplicates, i.e. tweets with the same or very close content, including retweets. For the cross-validation in Task 1, we ensured that duplicate tweets were not spread across different folds. In Task 4, this was achieved by removing all duplicate tweets from the training set after pre-processing and before training (i.e. for our experiments, numbers of unique tweets in Table 1 apply).

3 Experiments and System Descriptions

For Task 1, we experimented with two different systems, separately and in combination. The first system (labeled MTL) is a CNN+BiLSTM neural network with multi-task-learning (MTL) capabilities (Caruana, 1997). The multi-task architecture allows tackling multiple tasks (datasets) in a single model, based on the idea that complementary information from different tasks can lead to mutual benefit when they are trained jointly (see e.g. Crichton et al., 2017). The architecture distinguishes shared layers, where parameters are updated for all tasks during training, and task-specific layers with parameters dedicated to a single task. Our MTL architecture is able to handle different types of tasks, such as sequence labeling and document classification, in the same model. In the present configuration, the model was trained on data from Task 1, Task 2, and the CADEC corpus (Karimi et al., 2015), where the latter two served as helper tasks, solving the problem of span detection for ADRs. In the shared part of the model, character embeddings are combined with pre-trained word embeddings (Godin et al., 2015) into a bidirectional Long-Short Term Memory (BiLSTM) layer. In the task-specific layer, the sequence-labeling tasks are modeled with Conditional Random Fields (CRF), whereas the text-level classifier for Task 1 is based on the final state of the BiLSTM layer directly. Additionally, the Task-1 classifier uses a lexicon feature based on a fuzzy-match lookup in the MedDRA vocabulary.[2] We trained 10 different models in a cross-validation setting, using a held-out set to prevent overfitting through early stopping. The predicted labels are based on the mean of the scores of all folds (transformed by softmax).

We based the second system (labeled BERT) for Task 1 on BERT, a pre-trained language representation with a neural transformer architecture (Devlin et al., 2018). Our system merged parameters of 20 models (originating from 10-fold cross validation trained once for four epochs and once with early stopping[3]) into a single model (Utans, 1996; Junczys-Dowmunt et al., 2016). For this, we calculated the weighted sum of parameters across models: we weighted parameters of each model by their performance on the respective testing fold (measured as F-score and transformed by softmax). By applying this method, we first separately merged the systems resulting from training with early stopping and from training for 4 fixed epochs, and subsequently, merged the two resulting systems into a single system. For this last merging step, we gave the system resulting from merging early stopping systems nine times the weight of the other system which resulted from merging systems trained for a fixed number of epochs. For the last run (MTL+BERT), we combined predictions from all 20 BERT systems with the first system and a second MTL configuration which uses different word embeddings (Ellendorff et al., 2018) and omits lexicon features.

For Task 4, our submission consisted of three

[2]https://www.meddra.org/
[3]Early stopping was done on 0.2 of the training portion with a patience of 2.

	System	Precision	Recall	F-Score	Accuracy
Task 1	MTL	0.585	0.438	0.501	
	BERT	0.648	**0.567**	**0.605**	
	MTL+BERT	**0.705**	0.420	0.527	
	mean	*0.535*	*0.505*	*0.502*	
Task 4	Merge	0.839	**0.909**	**0.873**	**0.877**
	Average	0.988	0.614	0.757	0.818
	Join	**1.000**	0.515	0.680	0.775
	mean	*0.902*	*0.585*	*0.701*	*0.781*

Table 2: Official scores for our submissions, compared to mean scores of all participating systems (best results in bold).

		Precision	Recall	F-Score	Accuracy
Task 4	HC 1	0.910	0.988	0.947	0.944
	HC 2	0.706	0.774	0.739	0.754
	HC 3	0.750	0.790	0.769	0.839

Table 3: Official scores for Task 4, System 1 (Merged BERT models across contexts) by Health Context/Health Concern (HC).

different types of BERT-based ensemble systems. Our first system (labeled Merge) is similar to the second system (BERT) of Task 1. We trained two systems using 10-fold cross validation: one for infection and one for vaccination. Subsequently, we first merged the resulting systems across folds[4] and, in a second step, we merged the two resulting systems into one single system, giving nine times the weight to the system resulting from training on the infection dataset. This run has ranked first among all systems participating in the task. The second run (labeled Average) is again trained on both datasets separately using 10-fold cross validation, resulting in 20 independent systems. Labels are determined by averaging label probabilities returned by all 20 systems. Finally, the third run (Join) is trained on both datasets jointly but giving twice as much weight to all data points from the infection dataset, again using 10-fold cross validation, and probabilities were averaged across these 10 systems.

For both tasks, our BERT classifiers are based on the PyTorch implementation of BERT[5] and fine-tune the pre-trained model provided by Google research as *BERT-Base, Uncased*[6]. Where not mentioned otherwise, all systems were trained with the BertAdam optimizer for four epochs with

---

[4]For Task 4 we did not weight systems by their performance on the test fold, as we did for Task 1.

[5]https://github.com/huggingface/pytorch-pretrained-BERT

[6]https://github.com/google-research/bert

	System	Precision	Recall	F-Score
Task 1	Single	**0.765**	0.385	0.512
	Majority vote	0.688	0.462	0.552
	Merge unweighted	0.623	0.617	0.619
	Merge weighted	0.625	**0.621**	**0.623**

Table 4: Scores for post-submission runs for Task 1 (all BERT classifiers trained with early stopping). Single: single system trained on the whole training data; Majority vote: majority voting ensemble; Merge unweighted: unweighted parameter merging; Merge weighted: weighted parameter merging.

a batch size of 30 (Task 1) or 5 (Task 4), a learning rate of 5×10^{-5} and linear warmup schedule with a fixed number of 9050 training steps.

4 Results and Discussion

Table 2 shows official results on the test set. The official unlabeled test sets for Tasks 1 and 4 comprise 4575 and 285 tweets, respectively. Apart from an overall evaluation, systems submitted for Task 4 were also evaluated with respect to three different health contexts (also: health concerns), which were still undisclosed by the time we wrote this system description. For our best performing system (Merge), results for each health context can be found in Table 3.

In Task 1, the BERT-based model clearly outperformed our competing MTL-based approach. After the submission deadline, we used the evaluation interface to obtain test set evaluation scores for a BERT system, which for Task 1 only includes the systems trained with early stopping (i.e. we excluded the system which was trained for 4 fixed epochs). This still gave us a considerable improvement. Besides merging the 10 models into one, we also experimented with voting ensembles but found that merging models in fact gave us the best performance, with the weighted version still achieving a slight improvement compared to the unweighted version. Results for Task 1 post-submission runs can be found in Table 4.

Our results for both tasks show that merging models gives us a large improvement compared to traditional ensembling techniques (such as majority voting). Furthermore, merging parameters from several models into a single model means that only a single model is needed at prediction time. This brings a considerable advantage in terms of memory and computation time when predicting labels.

References

Rich Caruana. 1997. Multitask learning. *Machine Learning*, 28(1):41–75.

Gamal Crichton, Sampo Pyysalo, Billy Chiu, and Anna Korhonen. 2017. A neural network multi-task learning approach to biomedical named entity recognition. *BMC Bioinformatics*, 18(1):368.

Jacob Devlin, Ming-Wei Chang, Kenton Lee, and Kristina Toutanova. 2018. BERT: pre-training of deep bidirectional transformers for language understanding. *CoRR*, abs/1810.04805.

Tilia Ellendorff, Joseph Cornelius, Heath Gordon, Nicola Colic, and Fabio Rinaldi. 2018. UZH@SMM4H: System descriptions. In *Proceedings of the 2018 EMNLP Workshop SMM4H: The 3rd Social Media Mining for Health Applications Workshop and Shared Task*, pages 56–60, Brussels, Belgium.

Fréderic Godin, Baptist Vandersmissen, Wesley De Neve, and Rik Van de Walle. 2015. Multimedia Lab @ ACL W-NUT NER shared task: Named entity recognition for Twitter microposts using distributed word representations. In *Proceedings of the Workshop on Noisy User-generated Text*, pages 146–153, Beijing, China.

Xiaolei Huang, Michael Smith, Michael Paul, Dmytro Ryzhkov, Sandra Quinn, David Broniatowski, and Mark Dredze. 2017. Examining patterns of influenza vaccination in social media. In *AAAI Joint Workshop on Health Intelligence (W3PHIAI)*.

Marcin Junczys-Dowmunt, Tomasz Dwojak, and Rico Sennrich. 2016. The AMU-UEDIN submission to the WMT16 news translation task: Attention-based NMT models as feature functions in phrase-based SMT. In *Proceedings of the First Conference on Machine Translation*, pages 319–325, Berlin, Germany.

Sarvnaz Karimi, Alejandro Metke-Jimenez, Madonna Kemp, and Chen Wang. 2015. CADEC: A corpus of adverse drug event annotations. *Journal of Biomedical Informatics*, 55:73–81.

Alex Lamb, Michael J. Paul, and Mark Dredze. 2013. Separating fact from fear: Tracking flu infections on twitter. In *Proceedings of the 2013 Conference of the North American Chapter of the Association for Computational Linguistics: Human Language Technologies*, pages 789–795, Atlanta, Georgia. Association for Computational Linguistics.

Joachim Utans. 1996. Weight averaging for neural networks and local resampling schemes. In *Proc. AAAI-96 Workshop on Integrating Multiple Learned Models*, pages 133–138. AAAI Press.

Davy Weissenbacher, Abeed Sarker, Arjun Magge, Ashlynn Daughton, Karen O'Connor, Michael Paul, and Graciela Gonzalez-Hernandez. 2019. Overview of the fourth Social Media Mining for Health (SMM4H) shared task at ACL 2019. In *Proceedings of the 2019 ACL Workshop SMM4H: The 4th Social Media Mining for Health Applications Workshop & Shared Task*.

Identifying adverse drug events mentions in tweets using attentive, collocated, and aggregated medical representation

Xinyan Zhao,[1*] **Deahan Yu,**[1*] **V.G.Vinod Vydiswaran**[2,1]

[1] School of Information, [2] Department of Learning Health Sciences, University of Michigan
{zhaoxy,deahanyu,vgvinodv}@umich.edu
* denotes equal contribution

Abstract

Identifying mentions of medical concepts in social media is challenging because of high variability in free text. In this paper, we propose a novel neural network architecture, the Collocated LSTM with Attentive Pooling and Aggregated representation (**CLAPA**), that integrates a bidirectional LSTM model with attention and pooling strategy and utilizes the collocation information from training data to improve the representation of medical concepts. The collocation and aggregation layers improve the model performance on the task of identifying mentions of adverse drug events (ADE) in tweets. Using the dataset made available as part of the workshop shared task, we show that careful selection of neighborhood contexts can help uncover useful local information and improve the overall medical concept representation.

1 Introduction

Multiple studies have analyzed health forums and other social media for drug uses, pharmacovigilance, and effectiveness of medications (Nikfarjam et al., 2015; Daniulaityte et al., 2016). However, research related to drugs and adverse drug effects (ADE) in social media continues to grow rapidly. Automatically detecting ADE mentions in social media posts has been challenging due to the large variability of free text. One of the main challenges in studying natural language processing (NLP) approaches for medical information extraction is the lack of access to health-related information on social media (Weissenbacher et al., 2019).

Having a robust representation of words is important to train high-performance information extraction approaches. In domain-specific tasks, being able to properly represent domain words or concepts could significantly improve the mod-

els. While many studies have undertaken classifications of ADE mentions in posts with various state-of-the-art techniques (Nikfarjam et al., 2015; Weissenbacher et al., 2018), there is still room to improve for the task. For example, in many trained word embedding models (Pennington et al., 2014; Godin et al., 2015; Joulin et al., 2017), the embedding of each word is treated as a vector summarizing multiple semantic meanings for each word as independent dimensions. Indeed, pre-trained embeddings that are trained on a large data corpus usually provide robust representation for common words, compared to traditional feature-based techniques such as bag of words. Yet, for domain-specific tasks, a drawback of pre-trained embeddings is that representations of domain words may not be sufficiently tuned to be able to represent the expected meaning.

Attempts have been made previously to capture the word embedding for medical concepts from a variety of medical data sources (Huang et al., 2016). Similarly, domain-specific knowledge graphs have been shown effective as external resources for feature expansion to represent medical concepts (Choi et al., 2017; Wang et al., 2017). However, even domain-based knowledge graphs sometime contain redundant information stemming from how they are constructed (Yu et al., 2014; Paulheim, 2017; Zaveri et al., 2016). Following prior work by (Turenne, 2003) that show that co-occurring pattern of terms could be beneficial to classification tasks, in this work, we consider an alternate graph-based representation that utilizes local information derived from the training data set. We build a collocation graph – a word-based graph built from the training data set where nodes correspond to vocabulary words and edges between two nodes indicate the co-occurrence of the corresponding words. We investigate if a model built over the collocation graph could use

Proceedings of the 4th Social Media Mining for Health Applications (#SMM4H) Workshop & Shared Task, pages 62–70
Florence, Italy, August 2, 2019. ©2019 Association for Computational Linguistics

pre-trained word embeddings and other information to recognize medical concepts from data. We hypothesize that the representation of a medical word can be further enriched by its neighbors in the collocation graph.

In this paper, we propose Collocated **LSTM** with **A**ttentive **P**ooling and **A**ggregated representation (**CLAPA**), a novel approach that integrates bidirectional LSTM model with attention and pooling strategy and utilizes the collocation information in the training data set to help enhance the pre-trained word embedding of medical concepts. We show that our model leads to a significant improvement on an ADE detection task. To the best of our knowledge, this is the first attempt that utilizes local collocation information to improve the representation of domain concepts in social media.

To summarize, we make the following contributions in this paper:

- We propose a novel architecture that encodes locally stored domain information into sentence representation.

- Our work explores the possibility that limited training data could be better exploited by including attentive collocation information.

- We provide implication for other domain-related works where better representation of domain terms is important, especially when the data set is highly imbalanced.

2   Related work

Researchers have tackled the problem of identifying posts mentioning ADEs in social media in different ways. Various methods have been used in the 2018 Social Media Mining for Health Applications (SMM4H) shared task, ranging from statistical models such as support vector machines (SVM) to deep neural network models such as convolutional neural network (CNN), long short-term memory (LSTM), and bidirectional LSTM models. Fourteen teams participated in the 2018 SMM4H shared tasks (Weissenbacher et al., 2018), and used deep neural network models and various text processing steps such as correcting misspellings, accounting for class imbalance in data, and incorporating external resources. For the ADE mention classification task, the best system achieved an F1 score of 0.522, while the next best system achieved an F1 score of 0.478. The best system (Wu et al., 2018) was based on a bidirectional LSTM model with hierarchical tweet representation and multi-head self-attention.

In recent years, models such as CNN (Kim, 2014) and bidirectional LSTM (Graves and Schmidhuber, 2005) were used for text classification. In addition, models with attention mechanism, which incorporates information of other input tokens to improve representation of each token, was introduced by (Vaswani et al., 2017). Several max-pooling techniques, which help to detect important ngrams, were explored by (Jacovi et al., 2018) and (Zhou et al., 2016). Such mechanisms and technique have been powerful tools to build better text classification systems. To train distributed representations of words, (Mikolov et al., 2013) introduced Word2Vec in which each word is represented in a low-dimensional vector space. Other popular, pre-trained word embeddings include GloVe (Pennington et al., 2014), Word2vec over Twitter (Godin et al., 2015), and FastText (Joulin et al., 2017). Similarly, graph embedding techniques over large-scale networks were studied by numerous prior works, including LINE (Tang et al., 2015), DeepWalk (Perozzi et al., 2014), and Node2Vec (Grover and Leskovec, 2016). Although graph embedding is similar to word embedding, it is trained on not only nodes adjacent to each node but on the entire local network around the node. So, graph embedding could capture the relations between nodes, and has been used for multi-label classification and community detection (Grover and Leskovec, 2016; Qiu et al., 2018). Since most text-based graphs are typically reducible to a linear chain, and the ADE detection task is a binary classification problem, we focus on only the word embedding-based approaches in this paper.

3   Collocation and aggregated representation models

In this section, we describe the architecture of our model in detail. The model contains the following three key components — medical collocation embedding, sentence encoder, and max pooling. The overall architecture of our model is shown in Figure 1. For each word, the embedding is composed of two parts, namely, a pre-trained word embedding and an attentive neighborhood embedding. Attentive neighborhood embedding is de-

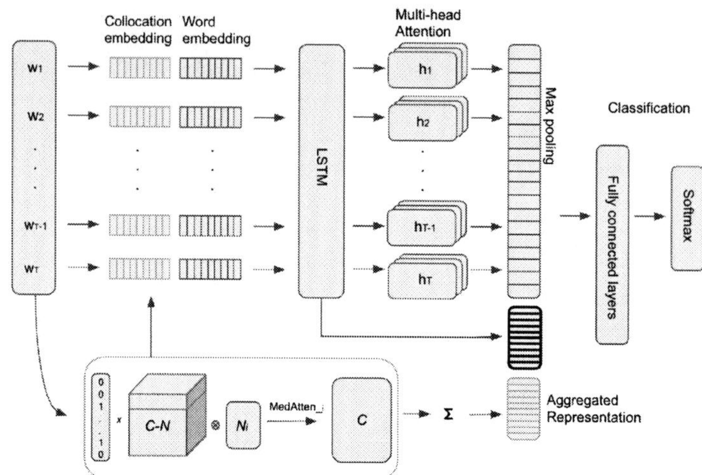

Figure 1: Overall architecture of the proposed model for identifying adverse drug events

rived from the Concept-Neighbor (C-N) tensor. In a C-N cube, each N_i represents the neighborhood for the i-th concept. Based on an attention vector ($MedAttn_i$), a concept embedding matrix C is formed in which c_i is the embedding for the concept. The collocation embedding for a word w_t will be c_i if w_t is the i-th concept, otherwise, the collocation embedding will be initialized to the zero vector. The concatenated embedding is then fed into an LSTM layer, and multi-head attention and maxpooling are applied to extract informative neurons, which are then concatenated with (1) the final state of the LSTM (sentence encoding) and (2) the sum of the concept embedding matrix. The final output is then computed via a fully connected neural network with a softmax function. Table 1 summarizes the notations used in this paper.

Notation	Definition
$W = [w_1, \dots, w_T]$	a sequence of words
$S = [s_1, \dots, s_{|S|}]$	medical concept set
$C = [c_1, \dots, c_{|S|}]$	concept matrix of size $R^{|S| \times d}$
$N_i = [n_{i1}, \dots, n_{iK}]$	neighborhood matrix of size $R^{K \times d}$ for the i-th concept.
C-N tensor	neighborhood tensor with the size $R^{|C| \times K \times d}$ composed by the neighborhood of each concept
w_t	t-th word in a text sequence.
s_i	i-th medical concept word in S
c_i	medical collocation embedding of the s_i
n_{ik}	word embedding of the k-th neighbor for the i-th concept in the concept set.
m_t	medical collocation embedding for the word w_t.
$|S|$	total number of concepts
T	total number of words in a sequence
K	maximum neighborhood size
L	total number of attention heads
d	dimension of word embedding
d_h	dimension of hidden states in LSTM

Table 1: Notation definitions

3.1 Medical collocation embedding

In order to better utilize the medical information embedded in text, we propose two word embedding methods – a pre-trained word embedding, and a second embedding method that enhances the pre-trained representation of medical terms by extracting information around those terms from the collocation graph.

Our medical collocation embedding can therefore be defined as following (Eq. 1):

$$MedAttn_j^i = \frac{exp(f(n_{ij}, W_1))}{\sum_k exp(f(n_{ik}, W_1))}$$
$$c_i = MedAttn_j^i \times N_i \tag{1}$$
$$m_t = \Delta(w_t, s_i) \times c_i$$

where $f(\cdot)$ represents a linear transformation and the $W_1^{K \times 1}$ is a trainable parameter matrix. $MedAttn_j^i$ calculates the attention that should be paid to the j-th neighbor for the concept s_i. Therefore, the embedding c_i is represented by the embedding of its neighborhood weighted by attention scores. Lastly, m_t represents the medical collocation embedding for the t-th word in text, w_t. If the word is matched to the i-th medical concept, then $m_t = c_i$. ($\Delta(x, y) = 1$ if $x = y$; 0 otherwise).

3.2 Aggregated Medical Representation

In addition to the word-based medical concept embedding described in Sec. 3.1, we propose another aggregated medical representation strategy using the collocation information that aggregates the medical concept information in a sentence into a fixed feature space.

First, we use an attentive embedding, c_i, described in Eq. 1, to construct a medical concept representation using the neighborhood information. Then, the aggregated representation is constructed, as follows:

$$c_i^* = c_i \oplus e(s_i)$$
$$Aggre = \sum_i \delta(i) \times c_i^* \qquad (2)$$

where $e(\cdot)$ is the function that retrieves the original representation of the medical concept word from pre-trained embedding. $\delta(\cdot) = 1$, when the sentence contains the concept word, and 0 otherwise. This aggregated medical representation serves as the residual medical information that is to be added to the output layer.

3.3 Sentence encoding

To encode a sentence for the classification task, we used an attention-based LSTM to encode the entire sentence into a fixed vector space. L attention heads are applied to re-represent hidden states. The new hidden states from the l-th attention head can be described as follows (Eq. 3):

$$H, s = \text{LSTM}([e(w_1) \oplus m_1, \dots, e(w_T) \oplus m_T])$$
$$SentAttn_t^l = \frac{exp(f(h_t, W_2^l))}{\sum_k exp(f(h_k, W_2^l))}$$
$$\hat{h}_t^l = SentAttn_t^l \cdot h_t$$
$$(3)$$

where $H = [h_1, \dots, h_T] \in R^{d_h \times T}$ is a hidden state matrix representing the information status at each time step, and d_h is a hidden dimension. $e(\cdot)$ and $f(\cdot)$ are the same as defined in Eq. 1. $SentAttn_t^l$ is a scalar representing the attention that should be

paid to h_t. Therefore, $\hat{h}_t^l$ is the attentive hidden state scaled by attention values in the l-th attention head.

3.4 Max pooling layer

Motivated by previous studies (Jacovi et al., 2018; Zhou et al., 2016), the application of max pooling behavior can highlight the important signals from features and hence improve classification tasks. Following these previous approaches, we apply a max pooling layer to extract important signals from the attentive hidden state in each attention head (Eq. 4).

$$signal_l = pooling(\hat{H}_l) \qquad (4)$$

where $\hat{H}_l = [\hat{h}_1^l, \dots, \hat{h}_T^l] \in R^{d_h \times T}$, and the pooling is applied on the dimension of d_h so that $signal_l \in R^{d_h}$ contains important signals from each hidden dimension.

3.5 Classification layer

In the final output layer, the classification decision is made on whether or not a sentence contains an ADE mention. A fully connected network module is implemented as:

$$r = s \oplus signal_1 \oplus \dots \oplus signal_L \oplus Aggre$$
$$r' = ReLU(U_1 r + b_1) \qquad (5)$$
$$\widehat{y} = softmax(U_2 r' + b_2)$$

where r is the combination of the final state of LSTM, multiple pooled states using max pooling, and aggregated medical concept representation. Each pooled state vector $signal_l$ comes from one attention layer (L attention layers in total) that is applied in sentence encoding (Eq. 3). U_1, U_2, b_1, and b_2 are parameters to be trained. Cross-entropy is used as the loss function for training:

$$loss = -\sum_i \sum_k y_k \log(\widehat{y}_k) \qquad (6)$$

4 Experiments

4.1 Data

For our experiments, we used the data set provided as part of Task 1 of the SMM4H 2019 shared tasks (Gonzalez-Hernandez et al., 2019). As summarized in Table 2, the total number of annotated tweets is 25,678. The data set was randomly split into a training set (80%) and a validation set

N = 25,678	Training set (80% of data)	Validation set (20% of data)
ADE tweets	1,892	485
Non-ADE tweets	18,650	4,651

Table 2: Number of ADE and non-ADE tweets in training and validation data sets.

(20%), while maintaining the target class proportions according to the original distribution. As a result, our training set contains 1,892 tweets that have an ADE mention (positive cases), and 18,650 tweets that do not have any mention of ADEs (negative cases). The validation set contains 485 positive and 4,651 negative tweets. We cleaned the tweets by separating punctuation marks, removing special characters, and replacing mentions, URLs, and number representations with normalized tokens. Finally, we used fastText (Joulin et al., 2017) as the pre-trained word embedding model.

4.2 Collocation graph

To build our collocation graph, we treat each unique word in the training set as a node, and add undirected edges from a word to adjacent words in a tweet. The collocation graph consists of 27,440 nodes and 188,329 edges. To reduce the graph size, we removed all words that appeared fewer than three times in the corpus. The resultant graph has 12,438 nodes and 159,759 edges. The mean of degree centrality is 25.39 (sd =114.59). 50% of the nodes have degrees less than 8, and 75% of the nodes have degrees less than 17.

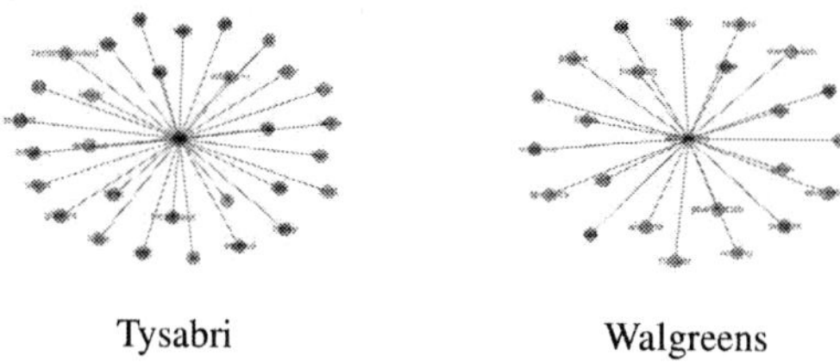

Figure 2: Examples of a collocation graph: *Tysabri* is considered as a medical concept while *Walgreens* is not considered as a medical concept.

Figure 2 shows the examples of a collocation graph. The graph has two colors: red and grey. The red nodes are words that are identified as medical concepts while the grey nodes are words that are not identified as medical concepts. The collocation graph on the left is for a medical word, *Tysabri*. The neighborhood of the word is comprised of both medical and non-medical words.

Tysabri contains other medical words as neighbors such as *infusion*, *treatment*, and *gilenya*. The collocation graph on the right is for a word, *walgreens*. It contains few medical words such as *cipro* and *miralax*.

4.3 Medical concepts extraction

MetaMap, a widely used system for identifying medical concepts in the unified language medical system (UMLS), is used to extract potential concepts from our tweet data set (Aronson, 2006). Given a sentence as input, MetaMap identifies phrases that could be medical concepts, and maps concepts to a preferred name using UMLS. However, since MetaMap is designed to parse clinical documents rather than free text on social media, we consider only those marked phrases that are the same as the preferred name as valid medical concepts. After processing, $1,340$ concepts were extracted by MetaMap from ADE tweets and $3,921$ concepts were extracted from non-ADE tweets. Concepts are later split into single words.

4.4 Training setup

All hyperparameters are jointly trained with a learning rate of 0.001 for ten epochs. In the experiments, we used FastText pretrained embedding, and the hidden size for LSTM is set to be 300. Number of multi-head attention layer is set to be 3. For each experiment, the score is taken from the average of five runs.

4.5 Results

To evaluate our model, we set two baselines: an attention-based LSTM model (Eq. 3), and an attention-based LSTM model with max pooling (Eq. 4). The results are presented in Table 3 as rows (1) and (4), respectively.

Model	Precision	Recall	F1
(1). LSTM+Attn (LA)	**0.6626**	0.4495	0.5356
(2). (1)+**colloc** (CLA)	0.6392	0.4639	0.5142
(3). (2)+**Aggr** (CLAA)	0.5181	**0.5918**	**0.5525**
(4). (1)+**Pool** (LAP)	**0.6475**	0.4887	0.5570
(5). (4)+**colloc** (CLAP)	0.6359	0.5546	0.5925
(6). (5)+**Aggr** (CLAPA)	0.6017	**0.5979**	**0.5998**
CLAPA on Test set	**0.5944**	**0.5431**	**0.5676**
Avg. system score	0.5351	0.5054	0.5019

Table 3: Comparison of models on Precision, Recall, and F1 measures for the ADE detection task on the validation set. The scores in the last two rows are over the test set of the 2019 SMM4H 2019 shared task 1.

As presented in Table 3, the model performance is significantly improved with the addition of collocation medical embedding and aggregated embedding, over the attention-based bi-direction LSTM models. Further, adding aggregated medical information helps improve recall, but reduces the model precision and only slightly increases the F1 score, compared to the collocation based model. Hence, while highlighting medical information can reduce false negative decisions, it also causes more instances to be labeled as ADE tweets, thereby increasing a false positive rate as well. The CLAPA model, that integrates both collocation and aggregated representation along with attentive pooling strategy performs the best.

When run against the test set for the shared task, the CLAPA model achieves a F1 score of 0.5676 (see Table 3). As a comparison, the average F1 score of systems participating in this task is 0.5019. This shows our CLAPA model performs significantly better than average on this task.

4.6 Model learning stability

To show that our model consistently works better even with smaller training data, we independently and randomly sampled 10%, 30%, 50%, 70%, and 90% data from training set and retrained the models. Figure 3 shows that our model consistently performed well on the validation set, even with reduced training size, compared to the baseline model of bidirectional LSTM model with attentive pooling (the "LAP" model). The results are similar to those on the full validation data set in Table 3, in that even when only a fraction of training data is available, the model achieves higher F1 score because of significantly better recall and at a relatively small reduction in precision.

4.7 Effect of concept vocabulary

Next, we analyzed the effect of medical concepts observed in the ADE tweets to understand if there is any difference in terms of the use of medical concepts in ADE tweets vs. non-ADE tweets. We calculated a propensity ratio of each medical term, based on number of times it appears in ADE tweets compared to non-ADE tweets. We found that *causing*, *gain*, *drowsiness*, and *sweats* are likely to appear in ADE tweets about 15 times more often than in non-ADE tweets. Similarly, *crippled* is likely to appear in an ADE tweet about 26 times more often than in a non-ADE tweet. Considering the highly skewed appearance ratio

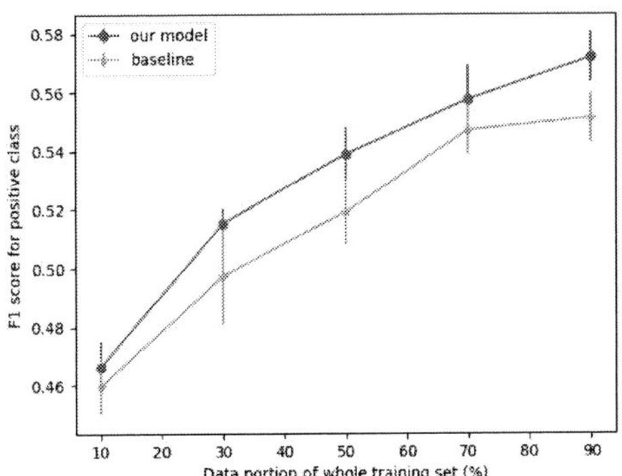

Figure 3: Effects of training size on model performance stability

for certain concepts, we analyzed the effect on using concepts from the ADE tweets alone. We compared two models – one trained over medical concepts identified from the ADE tweets and another trained over concepts from the entire training set, i.e. both ADE and non-ADE tweets.

Concepts from	Precision	Recall	F1
All tweets	0.6142	0.5546	0.5829
Only ADE tweets	0.6017	0.5979	**0.5998**

Table 4: Effects of concept vocabulary on model performance

As summarized in Table 4, the model trained with concepts from just the ADE tweets achieved a higher F1 score. While the precision is slightly lower, the model trained over concepts from ADE tweets has a significantly higher recall. On further analysis, we find that out of the $1,183$ concept words extracted from the ADE tweets, 866 concepts (73.2%) occurred more frequently in ADE tweets than in non-ADE tweets. However, when using the concepts words extracted from both ADE and non-ADE tweets, the number of concepts are higher ($n = 4,643$), but only $1,094$ concepts (23.6%) of those appear more frequently in the ADE tweets. This indicates that propensity ratio could be used for selecting medical concepts used in the ADE tweets as features.

4.8 Effects of neighborhood selection

We analyzed two additional questions related to parameter tuning:

(1) What method should be used to pick a neighbor? To answer this question, we fixed the neighborhood size as 15 words, and selected one of the following three methods to choose neighbors:

(a) Random: Given a node n, we randomly select k of its neighbors $n_1, n_2, \ldots, n_k \in N$, where N is a set of all neighbors for node n.

(b) Popularity: For each medical concept, we first selected a neighbor that has the highest degree. When node n_i has more neighbors than node n_j, we say that node n_i is more popular than node n_j. Then, given a node n, we select k popular neighbors $n_1, n_2, \ldots, n_k$ that have the highest degree. In case of ties in popularity, neighbors are selected at random from this set.

(c) Medical neighbor: Given node n, we add k medically-related neighbors.

For all three neighborhood selection methods, if the total number of first-degree neighbors is less than k, then an additional random selection is used among second-degree neighbors to fill the gap.

Table 5 shows the results using different selection methods under the two scenarios described in Section 4.7. The left column depicts the model trained on concepts from all tweets, and the right column represents the model trained with concepts from ADE tweets alone.

	F1 scores	
Selection method	**ADE+non-ADE**	**ADE**
Random	0.5796	0.5683
Popularity	0.5819	**0.5998**
Medical neighbor	**0.5829**	0.5887

Table 5: Effects of neighborhood selection methods on F1 scores on both ADE+non-ADE tweets and only ADE tweets

Table 5 shows that targeting at neighbors using either *popularity* or *medical* attributes always leads to better performance regardless of different scenarios. However, when using medical concepts of both ADE and non-ADE tweets, picking a medical neighborhood could be a better choice, whereas popular neighborhood is preferred when concepts are identified from ADE tweets. Medical neighborhood has a higher probability of including informative words related to ADE; and when only ADE tweets are considered, the frequency of co-occurrence of a neighbor and the concepts become more important. This explanation also aligns with how language models are usually trained.

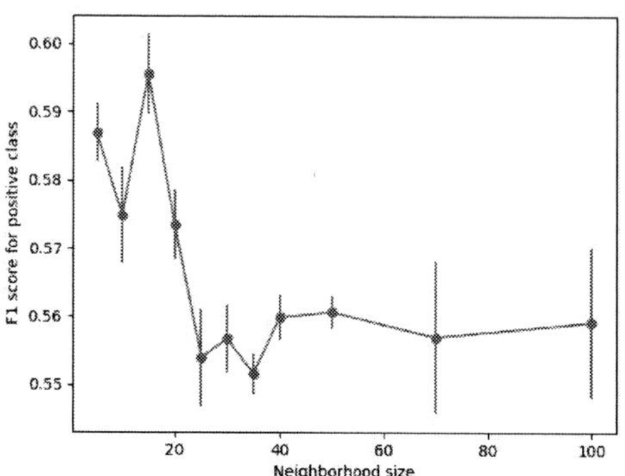

Figure 4: Effects of neighborhood size (k) on model performance

(2) How should we decide neighborhood size? We experimented with different neighborhood size. As shown in Fig. 4, as the neighborhood size k increases, the performance is not affected much when k is small (from 5 to 20). However, the performance drops significantly when k is larger ($k > 20$). We explain this by aligning back to our neighborhood selection method where we found that choosing good neighbors (popular or medically related) favors the model. We want to choose informative neighbors instead of all neighbors. Therefore, when k is small, the selected neighbors (high degree) can be easily differentiated from the ones not selected. However, when k is large, the selected neighbors become less informative because many unimportant, noisy, neighbor words (low degree/non-popular) may be included that harm the model.

5   Limitation and future work

After the above examination of our model, we argue that our model suffers from three main limitations. First, although MetaMap has been found useful at parsing medical notes, due to the different linguistic use on social media, running MetaMap on tweets may not identify relevant concepts. Second, the use of collocation graph and aggregated medical concept representation reduced precision of models, although the overall recall and F1 improved. Additional studies are need to further improve the precision. Third, the collocation graph is built solely on the training data set.

This may not favor the model when the data set is not representative enough to provide neighborhood of high quality. To address the first two issues, we believe a pre-trained state-of-the-art medication detection system could be helpful to identify high-quality medical concepts from tweets. For the third issue, we plan to use domain based knowledge base such as UMLS to expand the coverage of the limited data.

We used fastText as the pre-trained word embedding for our model. While fastText is trained on sub-word representations, models trained over medical or larger text corpora might provide additional contextual representation. Additional studies are needed to test our model on different pre-trained word embeddings such as Word2vec over Twitter (Godin et al., 2015). We also note that there is a difference in the use of medical related concepts in different classes by testing two scenarios — a model using medical concepts identified from both ADE and non-ADE cases and one using those from the ADE cases. In future, we plan to test this approach by exploring the use of unique nodes in different classes. Meanwhile, the application of our approach on other domain-specific tasks should be verified to examine the generalization of the approach.

6 Conclusion

In this work, we argue that a collocation graph can be utilized to enrich the representation of a medical concept. We further propose a novel neural network architecture that uses attentive information from a collocation graph to re-embed medical words. Our experiments show that, with a good selection of neighborhood, more useful local information can be accessed, which in turn improves the medical concept representation and the overall model performance in detecting mentions of adverse drug events in tweets.

Acknowledgment

We would like to thank Daniel Romero for his advice and feedback throughout the study.

References

Alan R Aronson. 2006. Metamap: Mapping text to the umls metathesaurus. *Bethesda, MD: NLM, NIH, DHHS*, pages 1–26.

Edward Choi, Mohammad Taha Bahadori, Le Song, Walter F. Stewart, and Jimeng Sun. 2017. Gram: Graph-based attention model for healthcare representation learning. In *KDD*.

Raminta Daniulaityte, Lu Chen, Francois R Lamy, Robert G Carlson, Krishnaprasad Thirunarayan, and Amit Sheth. 2016. "when 'bad' is 'good' ": Identifying personal communication and sentiment in drug-related tweets. *JMIR Public Health and Surveillance*, 2(2):e162.

Fréderic Godin, Baptist Vandersmissen, Wesley De Neve, and Rik Van de Walle. 2015. Multimedia lab @ acl wnut ner shared task: Named entity recognition for twitter microposts using distributed word representations. In *Proceedings of the Workshop on Noisy User-generated Text*, pages 146–153, Beijing, China. Association for Computational Linguistics.

Graciela Gonzalez-Hernandez, Davy Weissenbacher, Michael Paul, Abeed Sarker, Ari Z. Klein, Arjun Magge, Ashlynn R. Daughton, and Karen O'Connor. 2019. Social media mining for health applications (smm4h) workshop & shared task 2019. Https://healthlanguageprocessing.org/smm4h/.

Alex Graves and Jurgen Schmidhuber. 2005. Framewise phoneme classification with bidirectional lstm networks. In *Proceedings of the IEEE International Joint Conference on Neural Networks*, volume 4, pages 2047–2052.

Aditya Grover and Jure Leskovec. 2016. node2vec: Scalable feature learning for networks. In *Proceedings of the International Conference on Knowledge Discovery & Data Mining*, pages 855–864.

Jian Huang, Keyang Xu, and V.G. Vinod Vydiswaran. 2016. Analyzing multiple medical corpora using word embedding. In *IEEE International Conference on Healthcare Informatics (ICHI)*, volume 1, pages 527–533.

Alon Jacovi, Oren Sar Shalom, and Yoav Goldberg. 2018. Understanding convolutional neural networks for text classification. *arXiv preprint arXiv:1809.08037*.

Armand Joulin, Edouard Grave, Piotr Bojanowski, and Tomas Mikolov. 2017. Bag of tricks for efficient text classification. In *Proceedings of the 15th Conference of the European Chapter of the Association for Computational Linguistics: Volume 2, Short Papers*, pages 427–431. Association for Computational Linguistics.

Yoon Kim. 2014. Convolutional neural networks for sentence classification. In *Proceedings of the 2014 Conference on Empirical Methods in Natural Language Processing, EMNLP 2014, October 25-29, 2014, Doha, Qatar, A meeting of SIGDAT, a Special Interest Group of the ACL*, pages 1746–1751.

Tomas Mikolov, Ilya Sutskever, Kai Chen, Greg S Corrado, and Jeff Dean. 2013. Distributed representations of words and phrases and their compositionality. In *Advances in neural information processing systems*, pages 3111–3119.

Azadeh Nikfarjam, Abeed Sarker, Karen O'Connor, Rachel Ginn, and Graciela Gonzalez. 2015. Pharmacovigilance from social media: Mining adverse drug reaction mentions using sequence labeling with word embedding cluster features. *Journal of the American Medical Informatics Association : JAMIA*, 22(3):671–681.

Heiko Paulheim. 2017. Knowledge graph refinement: A survey of approaches and evaluation methods. *Semantic web*, 8(3):489–508.

Jeffrey Pennington, Richard Socher, and Christopher Manning. 2014. Glove: Global vectors for word representation. In *Proceedings of the 2014 conference on empirical methods in natural language processing (EMNLP)*, pages 1532–1543.

Bryan Perozzi, Rami Al-Rfou', and Steven Skiena. 2014. Deepwalk: online learning of social representations. In *KDD*.

Jiezhong Qiu, Yuxiao Dong, Hao Ma, Kuansan Wang, and Jie Tang. 2018. Network embedding as matrix factorization: Unifying deepwalk, line, pte, and node2vec. In *WSDM*.

Jian Tang, Meng Qu, Mingzhe Wang, Ming Zhang, Jun Yan, and Qiaozhu Mei. 2015. Line: Large-scale information network embedding. In *WWW*.

Nicolas Turenne. 2003. Learning semantic classes for improving email classification. In *Proceedings of Text Mining and Link Analysis Workshop*.

Ashish Vaswani, Noam Shazeer, Niki Parmar, Jakob Uszkoreit, Llion Jones, Aidan N Gomez, Ł ukasz Kaiser, and Illia Polosukhin. 2017. Attention is all you need. In I. Guyon, U. V. Luxburg, S. Bengio, H. Wallach, R. Fergus, S. Vishwanathan, and R. Garnett, editors, *Advances in Neural Information Processing Systems 30*, pages 5998–6008. Curran Associates, Inc.

Meng Wang, Mengyue Liu, Jun Liu, Sen Wang, Guodong Long, and Buyue Qian. 2017. Safe medicine recommendation via medical knowledge graph embedding. *CoRR*, abs/1710.05980.

Davy Weissenbacher, Abeed Sarker, Arjun Magge, Ashlynn Daughton, Karen O'Connor, Michael Paul, and Graciela Gonzalez-Hernandez. 2019. Overview of the fourth Social Media Mining for Health (SMM4H) shared task at ACL 2019. In *Proceedings of the 2019 ACL Workshop SMM4H: The 4th Social Media Mining for Health Applications Workshop & Shared Task*.

Davy Weissenbacher, Abeed Sarker, Michael J Paul, and Graciela Gonzalez-Hernandez. 2018. Overview of the third social media mining for health (smm4h) shared tasks at emnlp 2018. In *Proceedings of the 2018 EMNLP Workshop SMM4H: The 3rd Social Media Mining for Health Applications Workshop & Shared Task*, pages 13–16.

Chuhan Wu, Fangzhao Wu, Junxin Liu, Sixing Wu, Yongfeng Huang, and Xing Xie. 2018. Detecting tweets mentioning drug name and adverse drug reaction with hierarchical tweet representation and multi-head self-attention. In *Association for Computational Linguistics*, The 3rd Social Media Mining for Health Applications Workshop and Shared Task.

Dian Yu, Hongzhao Huang, Taylor Cassidy, Heng Ji, Chi Wang, Shi Zhi, Jiawei Han, Clare Voss, and Malik Magdon-Ismail. 2014. The wisdom of minority: Unsupervised slot filling validation based on multi-dimensional truth-finding. In *Proceedings of COLING 2014, the 25th International Conference on Computational Linguistics: Technical Papers*, pages 1567–1578.

Amrapali Zaveri, Anisa Rula, Andrea Maurino, Ricardo Pietrobon, Jens Lehmann, and Sören Auer. 2016. Quality assessment for linked data: A survey. *Semantic Web*, 7(1):63–93.

Peng Zhou, Zhenyu Qi, Suncong Zheng, Jiaming Xu, Hongyun Bao, and Bo Xu. 2016. Text classification improved by integrating bidirectional lstm with two-dimensional max pooling. *arXiv preprint arXiv:1611.06639*.

Correlating Twitter Language with Community-Level Health Outcomes

Arno Schneuwly
EPFL
arno.schneuwly@epfl.ch

Ralf Grubenmann
SpinningBytes
rg@spinningbytes.com

Séverine Rion Logean
Swiss Re
severine_rion@swissre.com

Mark Cieliebak
ZHAW
ciel@zhaw.ch

Martin Jaggi
EPFL
martin.jaggi@epfl.ch

Abstract

We study how language on social media is linked to diseases such as atherosclerotic heart disease (AHD), diabetes and various types of cancer. Our proposed model leverages state-of-the-art sentence embeddings, followed by a regression model and clustering, without the need of additional labelled data. It allows to predict community-level medical outcomes from language, and thereby potentially translate these to the individual level. The method is applicable to a wide range of target variables and allows us to discover known and potentially novel correlations of medical outcomes with life-style aspects and other socioeconomic risk factors.

1 Introduction

Surveys and empirical studies have long been a cornerstone of psychological, sociological and medical research, but each of these traditional methods pose challenges for researchers. They are time-consuming, costly, may introduce a bias or suffer from bad experiment design.

With the advent of big data and the increasing popularity of the internet and social media, larger amounts of data are now available to researchers than ever before. This offers strong promise new avenues of research using analytic procedures, obtaining a more fine-grained and at the same time broader picture of communities and populations as a whole (Salathé, 2018). Such methods allow for faster and more automated investigation of demographic variables. It has been shown that Twitter data can predict atherosclerotic heart-disease risk at the community level more accurately than traditional demographic data (Eichstaedt et al., 2015). The same method has also been used to capture and accurately predict patterns of excessive alcohol consumption (Curtis et al., 2018).

In this study, we utilize Twitter data to predict various health target variables (AHD, diabetes, various types of cancers) to see how well language patterns on social media reflect the geographic variations of those targets. Furthermore, we propose a new method to study social media content by characterizing disease-related correlations of language, by leveraging available demographic and disease information on the community level. In contrast to (Eichstaedt et al., 2015), our method is not relying on word-based topic models, but instead leverages modern state-of-the-art text representation methods, in particular sentence embeddings, which have been in increasing use in the Natural Language Processing, Information Retrieval and Text Analytics fields in the past years. We demonstrate that our approach helps capturing the semantic meaning of tweets as opposed to features merely based on word frequencies, which come with robustness problems (Brown and Coyne, 2018; Schwartz et al., 2018). We examine the effectiveness of sentence embeddings in modeling language correlates of the medical target variables (disease outcome).

Section 2 gives a generalized description of our method. We apply the previously described method to the tweets and health data in Section 3 The system's performance is evaluated in Section 4 followed by the discussion in Section 5. Our code is available on github.com/epfml/correlating-tweets.

2 Method

We are given a large quantity of text (sentences or tweets) in the form of social media messages by individuals. Each individual—and therefore each sentence—is assigned to a predefined category, for example a geographic region or a population subset. We assume the number of sentences to be sig-

Proceedings of the 4th Social Media Mining for Health Applications (#SMM4H) Workshop & Shared Task, pages 71–78
Florence, Italy, August 2, 2019. ©2019 Association for Computational Linguistics

nificantly larger than the number of communities. Furthermore, we assume that the target variable of interest, for example disease mortality or prevalence rate, is available for each community (but not for each individual). Our system consists of two subsystems:

1. *(Prediction)* The predictive subsystem makes predictions of target variables (e.g. AHD mortality rate) based on aggregated language features. The resulting linear predictions are applicable on the community level (e.g. counties) or on the individual level, and are trained using k-fold cross-validated Ridge regression.

2. *(Interpretability)* The averaged regression weights from the prediction system allow for interpretation of the system: We use a fixed clustering (which was obtained from all sentences without any target information), and then rank each topic cluster with respect to a prediction weight vector from point 1). The top and bottom ranked topic clusters for each target variable give insights into known and potentially novel correlations of topics with the target medical outcome.

In summary, the community association is used as a proxy or weak labelling to correlate individual language with community-level target variables. The following subsections give a more detailed description of the two subsystems.

2.1 System Description

Let S be the set of sentences (e.g. tweets), with their total number denoted as $|\mathcal{S}| = S$. Each sentence is associated to exactly one of the A communities $\mathcal{A} = \{a_1, \ldots, a_A\}$ (e.g. geographic regions). The function $\delta : \mathcal{S} \rightarrow \mathcal{A}$ defines this mapping. Let $\mathbf{y} \in \mathbb{R}^A$ be the target vector for an arbitrary target variable, so that each community a_j has a corresponding target value $y_{a_j} \in \mathbb{R}$.

Preprocessing and Embeddings. The complete linguistic preprocessing pipeline of a sentence is incorporated by the function $\rho(s_i)$, $\forall\ i \in \{1, \ldots, S\}$, which represents an arbitrary sentence s_i as a sequence of tokens. Each sentence s_i then is represented by a D-dimensional embedding vector providing a numerical representation of the semantics for the given short text:

$$\mathbf{x}_i = Sent2Vec(\rho(s_i)) \in \mathbb{R}^D. \qquad (1)$$

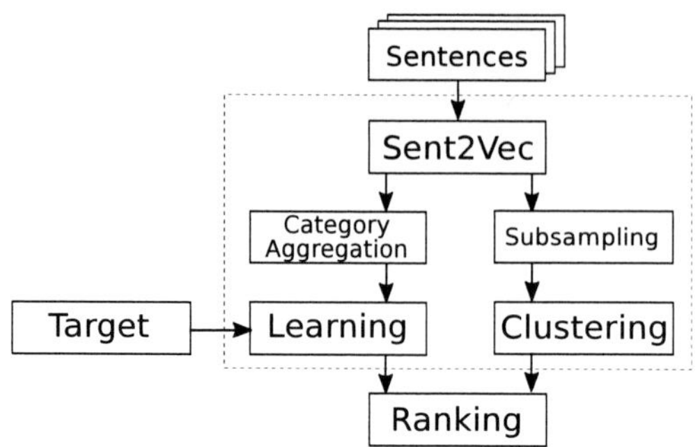

Figure 1: System Description.

While our method is generic for any text representation method, here Sent2Vec (Pagliardini et al., 2018) was chosen for its computational efficiency and scalability to large datasets.

2.2 Feature Aggregation

We use averaging of the sentence embedding vectors over each community to obtain the language features for each community. Formally, the complete feature matrix of all sentences is denoted as $\mathbf{X} \in \mathbb{R}^{S \times D}$. For our approach, the sentence embedding features are averaged over each community a_j. Formally, an individual feature $\overline{x}_{a_j,d}$ of the averaged embedding $\overline{\mathbf{x}}_{a_j} \in \mathbb{R}^{1 \times D}$ for a given community a_j is defined as

$$\overline{x}_{a_j,d} = \frac{1}{N_{a_j}} \sum_{x_i : s_i \in \mathcal{S}\ \wedge\ \delta(s_i) = a_j} x_{i,d}, \qquad (2)$$

where $N_{a_j} = |\{s_i : s_i \in S \wedge \delta(s_i) = a_j\}|$ is the number of sentences belonging to community a_j. Consequently, the aggregated community-level embedding matrix is given by

$$\overline{\mathbf{X}} = \begin{bmatrix} \overline{\mathbf{x}}_{a_1}^{\top} \\ \vdots \\ \overline{\mathbf{x}}_{a_A}^{\top} \end{bmatrix} \in \mathbb{R}^{A \times D}. \qquad (3)$$

2.3 Train-Test Split

Leveraging the targets available for each community, our regression method is applied to the aggregated features $\overline{\mathbf{X}}$ and the target $\mathbf{y}$. We employ K-fold cross-validation: the previously defined set $\mathcal{A}$ is split into K as equally sized pairwise disjoint subsets $\mathcal{A}_k$ as possible such that: $\mathcal{A} = \bigcup_{k=1}^{K} \mathcal{A}_k$, $\mathcal{A}_i \cap \mathcal{A}_j = \emptyset\ \forall i, j \in 1, \ldots, K, i \neq j$ and $|\mathcal{A}_1| \approx \cdots \approx |\mathcal{A}_K|$. The training set for a fold k is $\mathrm{TR}_k = \left(\bigcup_{i=1}^{K} \mathcal{A}_i \right) \setminus \mathcal{A}_k$ with the corresponding test set $\mathrm{TE}_k = \mathcal{A}_k$, where $N_k^{\theta} = |\mathrm{TR}_k|$ and $N_k^{\Lambda} =$

$|\mathrm{TE}_k|$. The operators $\theta_k : \{1, \ldots, N_k^\theta\} \to \mathrm{TR}_k$ and $\Lambda_k : \{1, \ldots, N_k^\Lambda\} \to \mathrm{TE}_k$ uniquely map the indexes to the corresponding communities a_j for the kth train-test split. For each split k the train and test embedding matrices respectively are defined as

$$\overline{\mathbf{X}}_{\theta_k} = \left[\overline{\mathbf{x}}_{\theta_k(1)}, \ldots, \overline{\mathbf{x}}_{\theta_k(N_k^\theta)}\right]^\top, \quad (4)$$

$$\overline{\mathbf{X}}_{\Lambda_k} = \left[\overline{\mathbf{x}}_{\Lambda_k(1)}, \ldots, \overline{\mathbf{x}}_{\Lambda_k(N_k^\Lambda)}\right]^\top. \quad (5)$$

Accordingly, we define the target vectors

$$\mathbf{y}_{\theta_k} = \left[y_{\theta_k(1)}, \ldots, y_{\theta_k(N_k^\theta)}\right]^\top, \quad (6)$$

$$\mathbf{y}_{\Lambda_k} = \left[y_{\Lambda_k(1)}, \ldots, y_{\Lambda_k(N_k^\Lambda)}\right]^\top. \quad (7)$$

2.4 Ridge Regression

For each train-test split k we perform linear regression from the community-level textual features $\overline{\mathbf{X}}_{\theta_k}$ to the health target variable $\mathbf{y}_{\theta_k}$. We employ Ridge regression (Hoerl and Kennard, 1970). In our context, the Ridge regression is defined as the following optimization problem:

$$\min_{\omega_k \in \mathbb{R}^D} \frac{1}{2A} \sum_{i=1}^{N_k^\theta} \left[y_{\theta_k(i)} - \overline{\mathbf{x}}_{\theta_k}^\top \omega_k\right]^2 + \lambda \|\omega_k\|_2^2, \quad (8)$$

where the optimal solution is

$$\omega_k^\star = \left(\overline{\mathbf{X}}_{\theta_k}^\top \overline{\mathbf{X}}_{\theta_k} + 2N_k^\theta \lambda \mathbf{I}\right)^{-1} \overline{\mathbf{X}}_{\theta_k}^\top \quad \in \mathbb{R}^D. \quad (9)$$

Within each each fold we tune the regularization parameter λ.

2.5 Prediction Subsystem

Let $\overline{\mathbf{y}}_{\Lambda_k} = \overline{\mathbf{X}}_{\Lambda_k} \omega_k^\star = [\overline{y}_{\Lambda_k(1)}, \ldots, \overline{y}_{\Lambda_k(N_k^\Lambda)}]^\top$ be the predicted values for the test set of the split k. The concatenated prediction vector for all splits is

$$\overline{\mathbf{y}}_\Lambda = \begin{bmatrix} \overline{\mathbf{y}}_{\Lambda_1}^\top \\ \vdots \\ \overline{\mathbf{y}}_{\Lambda_K}^\top \end{bmatrix} \in \mathbb{R}^A \quad (10)$$

Accordingly, we define the concatenated true target vector as

$$\mathbf{y}_\Lambda = \begin{bmatrix} \mathbf{y}_{\Lambda_1}^\top \\ \vdots \\ \mathbf{y}_{\Lambda_K}^\top \end{bmatrix} \in \mathbb{R}^A, \quad (11)$$

i.e., the set of individual scalars is identical to the entries in the original target vector $\mathbf{y}$. The predictive performance of the system can be assessed through the following metrics:

- Pearson Correlation Coefficient

- Mean Average Error of prediction (MAE)

- Classification Accuracy for Quantile Prediction

The first two metrics are evaluated with the vectors $\overline{\mathbf{y}}_\Lambda$ and $\mathbf{y}_\Lambda$ from all folds. In the quantile-based assessment we independently bin the true values $\mathbf{y}_\Lambda$ and the predicted values $\overline{\mathbf{y}}_\Lambda$ into C different quantiles. Each individual true and predicted value is assigned to a quantile $c_j \in \{c_1, \ldots, c_C\}$. These assignments can be used to visually compare results on a heat-map or as regular evaluation scores in terms of accuracy.

2.5.1 Ridge-Weight Aggregation

For the final prediction model, the regression weights $\omega_k^\star$ from Ridge regression are averaged over the K folds, i.e. $\overline{\omega} = \frac{1}{K} \sum_{k=1}^K \omega_k^\star$.

For every sentence embedding $\mathbf{x}_q$, the prediction is computed as $\overline{y}_q = \mathbf{x}_q^\top \overline{\omega} \in \mathbb{R}$.

2.6 Interpretation Subsystem: Cluster Ranking

We employ predefined textual topic clusters—which are independent of any target values—in order to enable interpretation of the textual correlates. Each cluster is a collection of sentences and should, intuitively, be interpretable as a topic, e.g. separate topics about indoor and outdoor activities as shown in Fig. 4. For each cluster m a ranking score can be computed with respect to a linear prediction model $\overline{\omega}$ such as defined above. Let $\mathcal{Q}_m = \{q : \zeta(q) = m \land q \in \mathcal{Q}\}$ be the set of sentences assigned to cluster m. The score ι_m for the cluster m is the average of all predictions $\overline{y}_q = \mathbf{x}_q^\top \overline{\omega}$ within the cluster m:

$$\iota_m = \frac{1}{|\mathcal{Q}_m|} \sum_{\overline{y}_q : \, q \in \mathcal{Q}_m} \overline{y}_q \quad (12)$$

By ordering the scores ι_m of all clusters, we obtain the final ranking sequence of all clusters, with respect to the target-specific model $\overline{\omega}$.

Clustering Preprocessing. For obtaining the fixed clustering, as $\mathbf{X}$ is a very large matrix, clustering might require subsampling to reduce computational complexity. Hence, Q out of the S embeddings in $\mathcal{S}$ are randomly subsampled into the set $\mathcal{Q}$. The mapping $\Phi(\mathcal{Q}) = [\phi(1), \ldots, \phi(Q)]^\top$

is a uniformly random selection of row indexes in $\mathbf{X}$ out of $\binom{N}{Q}$. We define the subsampled data matrix as $\mathbf{X}_Q = \left[\mathbf{x}_{\phi(1)}, \ldots, \mathbf{x}_{\phi(Q)}\right]^\top \in \mathbb{R}^{Q \times D}$.

The subset $\mathbf{X}_Q$ is clustered with the Yinyang K-Means algorithm (Ding et al., 2015). We use M centroids and the cosine similarity as a distance function. The cluster assignment vector $\mathbf{M} \in [1, \ldots, M]^\top$ assigns one cluster for each embedding in $\mathbf{X}_Q$. Accordingly, the operator $\zeta : \{1, \ldots, Q\} \rightarrow \{1, \ldots, M\}$ indicates the assigned cluster m for a given sentence s in $\mathcal{Q}$ (see cluster ranking above). The cluster centers are defined in $\mathbf{M}_Q \in \mathbb{R}^{Q \times D}$.

3 Data sources

We apply the method described in Section 2 to the following setting: The pool of sentences $\mathcal{S}$ consists of geotagged Tweets. The assigned locations are in the United States. The geotags are categorized into US-counties which represent the set of communities $\mathcal{A}$. The target variables y are health-related variables, for example normalized mortality or prevalence rates. We focus on cancer and AHD mortality as well as on diabetes prevalence. Hence, the quantile-based predictions give a categorization of the Ridge regression predictions on a US-county level. The ranked topics assess what language might relate to higher or lower rates of the corresponding disease. Table 1 provides an overview of the size of the data sources, the year the data was collected in and the mean μ and standard deviation σ of the target variables. Not all counties are covered in the publicly available datasets, usually being limited to more populous counties. The collected Tweets are from 2014 and 2015. The target variables are the union-averaged values from 2014 and 2015: if the target variable is available for both years the two values are averaged. Conversely, if a county data point is only available for one, but not both years, we use this standalone value.

3.1 Datorium Tweets

Tweets are short messages of no more than 140 characters[1] published by users of the Twitter platform. They reflect discussions, thoughts and activities of its users. We use a dataset of approximately 144 million tweets collected from first of June 2014 to first of June 2015 (Datorium, 2017).

Name	# tweets	Year	
Datorium	147M	14/15	

Name	# counties	Year	μ, σ
AHD	803	14/15	43.0, 16.1
Diabetes	3129	13	9.7, 2.2
Breast	487	13/14	12.4, 2.8
Colon	490	13/14	12.1, 3.0
Liver	293	13/14	7.5, 2.4
Lung	1612	13/14	52.4, 16.2
Melanoma	162	13/14	3.8, 1.2
Prostate	351	13/14	8.5, 2.0
Stomach	136	13/14	3.6, 0.9

Table 1: Overview of data sources.

Each tweet was geotagged by the submitting user with exact GPS coordinates and all tweets are from within the US, allowing accurate county-level mapping of individual tweets.

3.2 AHD & Cancer Mortality

Our source of the statistical county-level target variables is the CDC WONDER[2] database (CDC, 2018) for AHD and cancer. Values are given as deaths per capita (100'000).

3.3 Diabetes Prevalence

We use county-wise age-adjusted diabetes prevalence data from the year 2013 (CDC, 2016), provided as percent of the population afflicted with type II diabetes. The data is available for almost all the 3144 US counties, making it a valuable target to use.

4 Results

The results of our method for the various target variables are listed in Table 2 along with the performance of the baseline model outlined in Section 4.1. We provide the Pearson correlation (ρ) and the mean absolute error (MAE) of our system along with the baseline model's Pearson correlation.

4.1 LDA Baseline Model

We reimplemented the approach proposed by Eichstaedt et al. (2015) as a baseline for comparison, and were able to reproduce their findings about AHD with recent data: similar results were

---

[1]Twitter increased the limit to 280 characters in 2017, which doesn't affect our data.

[2]US Centers for Disease Control and Prevention - Wide-ranging Online Data for Epidemiologic Research.

found with the Datorium Twitter dataset (Datorium, 2017) and CDC AHD data from 2014 and 2015. Their approach averages topics generated with Latent Dirichlet Allocation (LDA) of tweets per county as features for Ridge regression. We do not use any hand-curated emotion-specific dictionaries, as these did not impact performance in our experiments. We used the predefined *Facebook LDA* coefficients of Eichstaedt et al. (2015), updated them with the word frequencies of our collected Twitter data (Datorium, 2017). Our results are computed with a 10-fold cross-validation and without any feature selection.

Type	ρ	ρ *LDA*	*MAE*
AHD	**0.46**	0.31	13.4
Diabetes	**0.73**	0.72	1.1
Breast	**0.44**	0.42	1.80
Colon	**0.55**	0.51	1.87
Liver	0.29	**0.40**	1.59
Lung	**0.68**	0.63	8.44
Melanoma	**0.72**	0.61	0.68
Prostate	**0.39**	0.38	1.34
Stomach	0.44	**0.51**	0.72

Table 2: Results of predictions on different health targets. ρ: our system (Section 2.5), ρ LDA: topic model baseline (Eichstaedt et al. (2015), Section 4.1), MAE: mean absolute error of our system (Section 2.5).

4.2 Detailed Results

In this section we discuss a selection of our results in detail, with additional information available in Appendix A.1.

Diabetes has a strong demographic bias, with a higher prevalence in the south-east of the US, the so called *diabetes belt*. Compared to the national average, the african-american population in the diabetes belt has a higher risk of diabetes by a factor of more than 2 (Barker et al., 2011) and the southeast of the US has a large african-american population. Therefore, linguistic features (Green, 2002) common in african-american are a strong predictor of diabetes rates. The model learns these linguistic features, as seen in Figure 3, and its predictions closely match the actual geographic distribution, as seen in Figure 2. A moderate alcohol consumption is linked to a low risk of type II diabetes compared to no or excessive consumption (Koppes et al., 2005). The strongest negatively correlated word clouds in Figure 3 support this finding.

The most positively related word clouds for melanoma in Figure 4 are related to outdoor activities (Elwood et al., 1985). Conversely, the strongest negatively correlated word clouds suggest indoor activity related language.

5 Discussion

In this paper, we introduced a novel approach for language-based predictions and correlation of community-level health variables. For various health-related demographic variables, our approach outperforms in most cases (Table 2) similar models based on traditional demographic data by using only geolocated tweets. Our approach provides a method for discovering novel correlations between open-vocabulary topics and health variables, allowing researchers to discover yet unknown contributing factors based on large collections of data with minimal effort.

Our findings, when applying our method to AHD risk, diabetes prevalence and the risk of various types of cancers, using geolocated tweets from the US only, show that a large variety of health-related variables can be predicted with surprisingly high precision based solely on social media data. Furthermore, we show that our model identifies known and novel risk or protective factors in the form of topics. Both aspects are of interest to researchers and policy makers. Our model proved to be robust for the majority of targets it was applied to.

For AHD risk, we show that our approach significantly outperforms previous models based on topic models such as LDA or traditional statistical models (Eichstaedt et al., 2015), achieving a ρ-value of 0.46, an increase of 0.09 over previous approaches. For diabetes prevalence our model correctly predicts its geographic distribution by identifying linguistic features common in high-prevalence areas among other features, with a ρ-value of 0.73. For melanoma risk, it finds a high-correlation with the popularity of outdoor activities, corresponding to exposure to sunlight being one of the main risk factors in skin cancer, with an overall ρ-value of 0.72.

One of the main limitations of our approach is the need for a large collection of sentences for each community as well as a large number of communities with target variables, leading to potentially unreliable results when this is not the case, such as for social media posts by individuals

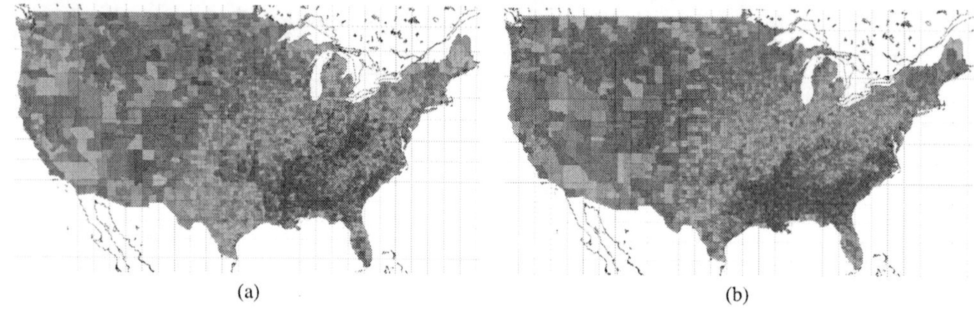

(a)   (b)

Figure 2: Quantiles of the prevalence of **diabetes**. (a) Target values (b) Predicted values from tweets

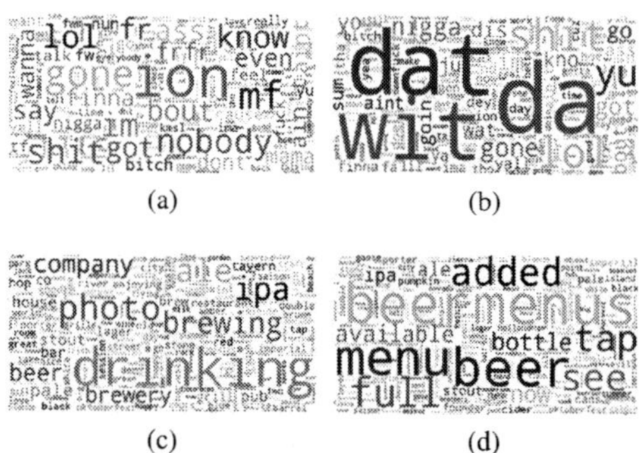

Figure 3: Word clouds of topics correlating with **diabetes**: (a) (b) strongest positive correlation (c) (d) strongest negative correlation among $M = 2000$ clusters.

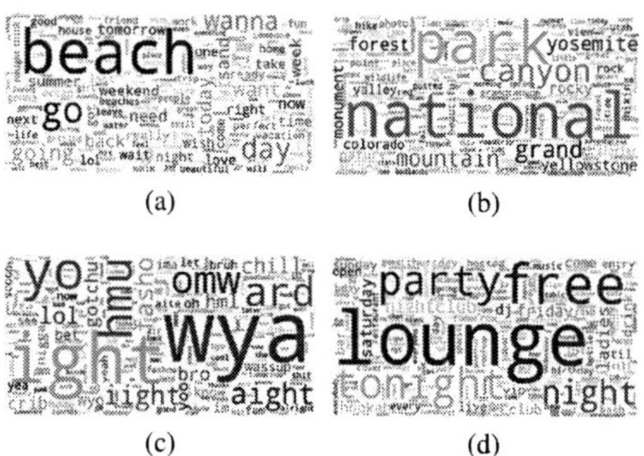

Figure 4: Word clouds of topics correlating with **melanoma**: (a) (b) strongest positive correlation (c) (d) strongest negative correlation among $M = 2000$ clusters.

or when modeling target values which are only available in e.g. few counties. Further research is needed to ascertain whether significant results can also be achieved in such scenarios, and if robustness of our approach is improved compared to bag-of-words-based baselines (Eichstaedt et al., 2015; Brown and Coyne, 2018; Schwartz et al., 2018). Furthermore, all mentioned approaches rely on *correlation*, and thus do not provide a way to determine any *causation*, or ruling out of potential underlying factors not captured by the model. Even though using social media data introduces a non-negligible bias towards users of social media, our approach was able to predict target variables tied to very different age-groups, which is encouraging and supports the robustness of our approach.

Our method captures language features on a community scale. This raises the question of how these findings can be translated to the individual person. Theoretically, a community-based model as described above could be used to rank social media posts or messages of an individual user, with respect to specific health risks. However, as we currently do not have ground truth values on the individual level, and since user's social media history has very high variance, this is left for future investigation.

Future research should also address the applicability of our model to textual data other than Twitter and potentially from non-social media sources, to communities that are not geography based, to the time evolution of topics and health/lifestyle statistics, as well as to targets that are not health related. The general methodology offers promise for new avenues for data-driven discovery in fields such as medicine, sociology and psychology.

Acknowledgements. We would like to thank Ahmed Kulovic and Maxime Delisle for valuable input and discussions.

References

Lawrence E. Barker, Karen A. Kirtland, Edward W. Gregg, Linda S. Geiss, and Theodore J. Thompson. 2011. Geographic distribution of diagnosed diabetes in the us: a diabetes belt. *American journal of preventive medicine*, 40(4):434–439.

Nicholas JL. Brown and James C. Coyne. 2018. Does Twitter language reliably predict heart disease? a commentary on eichstaedt et al.(2015a). *PeerJ*, 6:e5656.

CDC. 2016. County data. *National Center for Chronic Disease Prevention and Health Promotion, Division of Diabetes Translation*.

CDC. 2018. CDC WONDER. *WONDER – Wide-ranging Online Data for Epidemiologic Research*.

Brenda Curtis, Salvatore Giorgi, Anneke EK. Buffone, Lyle H. Ungar, Robert D. Ashford, Jessie Hemmons, Dan Summers, Casey Hamilton, and H. Andrew Schwartz. 2018. Can Twitter be used to predict county excessive alcohol consumption rates? *PloS one*, 13(4):e0194290.

Datorium. Geotagged Twitter posts from the united states: A tweet collection to investigate representativeness [online]. 2017.

Yufei Ding, Yue Zhao, Xipeng Shen, Madanlal Musuvathi, and Todd Mytkowicz. 2015. Yinyang K-means: A drop-in replacement of the classic K-means with consistent speedup. In *ICML'15 - Proceedings of the 32nd International Conference on International Conference on Machine Learning*.

Johannes C. Eichstaedt, Hansen Andrew Schwartz, Margaret L. Kern, Gregory Park, Darwin R. Labarthe, Raina M. Merchant, Sneha Jha, Megha Agrawal, Lukasz A. Dziurzynski, Maarten Sap, et al. 2015. Psychological language on Twitter predicts county-level heart disease mortality. *Psychological science*, 26(2):159–169.

J. Mark Elwood, Richard P. Gallagher, GB. Hill, and JCG. Pearson. 1985. Cutaneous melanoma in relation to intermittent and constant sun exposurethe western canada melanoma study. *International journal of cancer*, 35(4):427–433.

Lisa J. Green. 2002. *African American English: a linguistic introduction*. Cambridge University Press.

Arthur E. Hoerl and Robert W. Kennard. 1970. Ridge regression: Biased estimation for nonorthogonal problems. *Technometrics*, 12(1):55–67.

Lando LJ. Koppes, Jacqueline M. Dekker, Henk FJ. Hendriks, Lex M. Bouter, and Robert J. Heine. 2005. Moderate alcohol consumption lowers the risk of type 2 diabetes: a meta-analysis of prospective observational studies. *Diabetes care*, 28(3):719–725.

Matteo Pagliardini, Prakhar Gupta, and Martin Jaggi. 2018. Unsupervised Learning of Sentence Embeddings using Compositional n-Gram Features. In *NAACL 2018 - Conference of the North American Chapter of the Association for Computational Linguistics*.

Marcel Salathé. 2018. Digital epidemiology: what is it, and where is it going? *Life sciences, society and policy*, 14(1):1.

H. Andrew Schwartz, Salvatore Giorgi, Margaret L. Kern, Gregory Park, Maarten Sap, Darwin R. Labarthe, Emily E. Larson, Martin Seligman, Lyle H. Ungar, et al. 2018. More evidence that Twitter language predicts heart disease: a response and replication.

A.1   Additional Figures

(a)

(b)

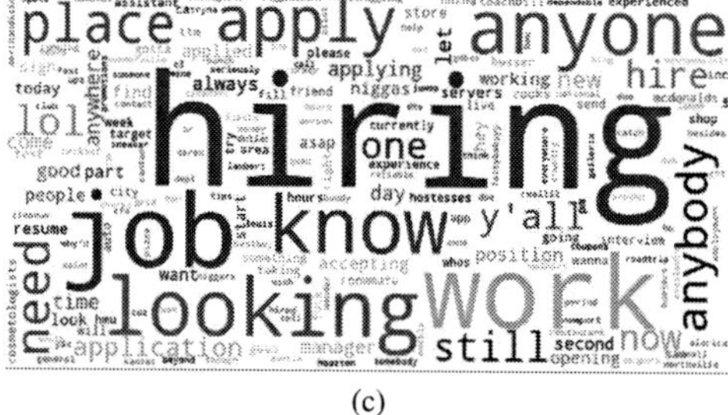

(c)

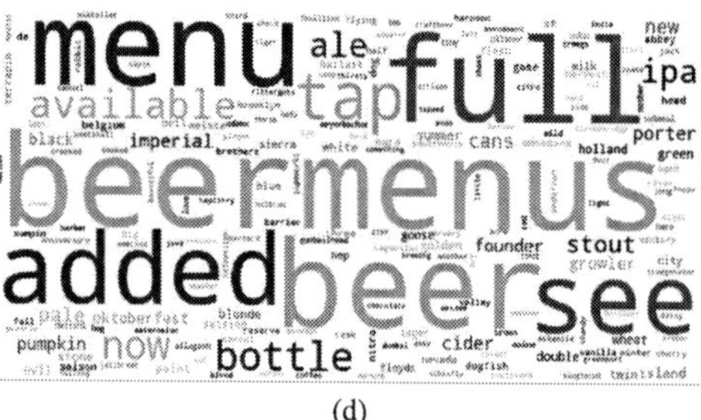

(d)

Figure 5: Word clouds of topics correlating with **colorectal cancer**: (a) (b)strongest positively correlated topics (c) (d) strongest negatively correlated topics among $M = 2000$ clusters.

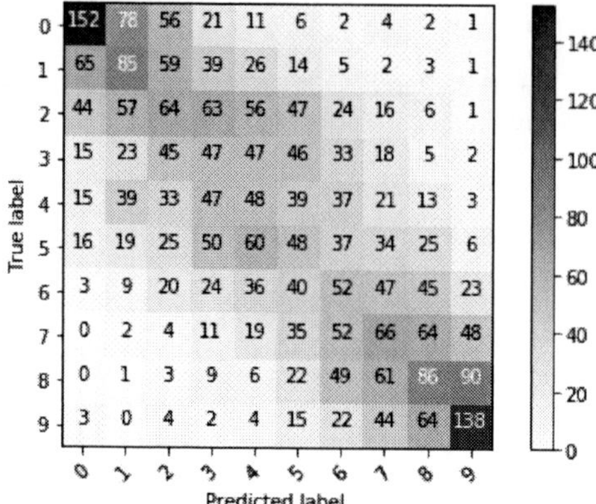

Figure 6: Confusion matrix for decile-based prediction of **diabetes** prevalence.

A.2   Implementation Details

Tweets were collected according to the provided datorium IDs using the Tweepy[3] library. The tweets were then imported into Google BigQuery[4] and processed using Apache Beam[5]. The sentence embeddings were computed using the official Sent2Vec source code and the provided 700-dimensional pre-trained model for tweets (using bigrams)[6]. Clustering was performed by libKM-CUDA[7]. Scikit-learn[8] was used for 10-fold cross validation, Ridge regression, calculating the correlation and hyperparameter search.

---

[3] https://www.tweepy.org/
[4] https://cloud.google.com/bigquery/
[5] https://beam.apache.org/
[6] https://github.com/epfml/sent2vec
[7] https://github.com/src-d/kmcuda
[8] https://scikit-learn.org/stable/

Affective Behaviour Analysis of On-line User Interactions:
Are On-line Support Groups more Therapeutic than Twitter?

Giuliano Tortoreto[†]**, Evgeny A. Stepanov**[†]**, Alessandra Cervone**[†]**,**
Mateusz Dubiel[⋆]**, Giuseppe Riccardi**[†]

[†]Signals and Interactive Systems Lab, University of Trento, Italy
[⋆]University of Strathclyde, Glasgow, UK
`name.surname@unitn.it, name.surname@strath.ac.uk`

Abstract

The increase in the prevalence of mental health problems has coincided with a growing popularity of health related social networking sites. Regardless of their therapeutic potential, on-line support groups (OSGs) can also have negative effects on patients. In this work we propose a novel methodology to automatically verify the presence of therapeutic factors in social networking websites by using Natural Language Processing (NLP) techniques. The methodology is evaluated on on-line asynchronous multi-party conversations collected from an OSG and Twitter. The results of the analysis indicate that therapeutic factors occur more frequently in OSG conversations than in Twitter conversations. Moreover, the analysis of OSG conversations reveals that the users of that platform are supportive, and interactions are likely to lead to the improvement of their emotional state. We believe that our method provides a stepping stone towards automatic analysis of emotional states of users of online platforms. Possible applications of the method include provision of guidelines that highlight potential implications of using such platforms on users' mental health, and/or support in the analysis of their impact on specific individuals.

1 Introduction

Recently, people have started looking at online forums either as a primary or secondary source of counseling services (Vogel et al., 2007). McMahon (2016) reported that over the first five years of operation (2011-2016), ReachOut.com – Ireland's online youth mental health service – 62% of young people would visit a website for support when going through a tough time. With the expansion of the Internet, there has been a substantial growth in the number of users looking for psychological support online.

The importance of the on-line life of patients has been recognized in research as well. Amichai-Hamburger et al. (2014) stated that the online life

of patients constitutes a major influence on their self-definition. Furthermore, according to Back et al. (2010), the social networking activities of an individual, offer an important reflection of their personality. While dealing with patients suffering from psychological problems, it is important that therapists do not ignore this pivotal source of information which can provide deep insights into their patients' mental conditions.

Acceptance of on-line support groups (OSG) by Mental Health Professionals is still not established (Andersson, 2017). Since OSG can have double-edged effects on patients and the presence of professionals is often limited, we argue that their properties should be further studied. According to Barak et al. (2008) OSG effectiveness is hard to assess, while some studies showed OSG's potential to change participants' attitudes, no such effect was observed in other studies (see Related Work Section for more details). Furthermore the scope of previous work on analysis of users' behaviour in OSG has been limited by the fact that they relied on expert annotation of posts and comments (Mayfield et al., 2012).

We present a novel approach for automatically analysing online conversations for the presence of therapeutic factors of group therapy defined by Yalom and Leszcz (2005) as "the actual mechanisms of effecting change in the patient". The authors have identified 11 therapeutic factors in group therapy: Universality, Altruism, Instillation of Hope, Guidance, Imparting information, Developing social skills, Interpersonal learning, Cohesion, Catharsis, Existential factors, Imitative behavior and Corrective recapitulation of family of origin issues. In this paper, we focus on 3 therapeutic factors: Universality, Altruism and Instillation of Hope (listed below), as we believe that these can be approximated by using established NLP techniques (e.g. Sentiment Analysis, Dialogue Act tagging etc.).

Proceedings of the 4th Social Media Mining for Health Applications (#SMM4H) Workshop & Shared Task, pages 79–88
Florence, Italy, August 2, 2019. ©2019 Association for Computational Linguistics

1. Universality: the disconfirmation of a user's feelings of uniqueness of their mental health condition.

2. Altruism: others offer support, reassurance, suggestions and insight.

3. Instillation of Hope: inspiration provided to participants by their peers.

The selected therapeutic factors are analysed in terms of illocutionary force[1] and attitude[2]. Due to the multi-party and asynchronous nature of on-line social media conversations, prior to the analysis, we extract conversation threads among users – an essential prerequisite for any kind of higher-level dialogue analysis (Elsner and Charniak, 2010). Afterwards, the illocutionary force is identified using Dialogue Act tagging, whereas the attitude by using Sentiment Analysis. The quantitative analysis is then performed on these processed conversations.

Ideally, the analysis would require experts to annotate each post and comment on the presence of therapeutic factors. However, due to time and cost demands of this task, it is feasible to analyse only a small fraction of the available data. Compared to previous studies (e.g. (Mayfield et al., 2012)) that analysed few tens of conversations and several thousand lines of chat; using the proposed approach – application of Dialogue Acts and Sentiment Analysis – we were able to automatically analyse approximately 300 thousands conversations (roughly 1.5 million comments).

The rest of the paper is structured as follows. In Section 2 we introduce related work. Next, in Section 3 we describe the pre-processing pipeline and the methodology to perform thread extraction on asynchronous multi-party conversations. In Section 4 we provide the describe the final dataset used for the analysis, and in Section 5 we present the results of our analysis. Finally, in Section 6 we provide concluding remarks and future research directions.

---

[1]The illocutionary force of an utterance is the speaker's intention in producing that utterance according to Loos (2003).

[2]"The attitude may be either his or her affective state, namely the emotional state of the author when writing, or the intended emotional communication, namely the emotional effect the author wishes to have on the reader" Gala et al. (2014).

2 Related Work

On-line support groups have been analyzed for various factors before. For instance, Chung (2013) analysed stress reduction in on-line support group chat-rooms, and the effects of on-line social interactions. Such studies mostly relied on questionnaires and were based on a small number of users. Nevertheless, in Chung (2013), the author showed that social support facilitates coping with distress, improves mood and expedites recovery from it. These findings highlight that, overall, on-line discussion boards appear to be therapeutic and constructive for individuals suffering alcohol-abuse.

Application of NLP to the analysis of mental health-related conversation has been studied as well (e.g. (Ghosh et al., 2017; Stepanov et al., 2018)). Mayfield et al. (2012) applied sentiment-analysis combined with extensive turn-level annotation to investigate stress reduction in on-line support group chat-rooms, showing that sentiment-analysis is a good predictor of entrance stress level. Furthermore, similar to our setting, they applied automatic thread-extraction to determine conversation threads.

Kissane et al. (2007) have shown that on-line support group therapy increased the quality of life of patients with metastatic breast cancer. Since many original posters reported the benefits of group therapy on patients (McDermut et al., 2001; Amichai-Hamburger et al., 2014; Tartakovsky, 2016; Espie et al., 2012; Gary and Remolino, 2000; Yalom and Leszcz, 2005), we evaluate the effect of the user interaction using sentiment scores of comments in on-line support groups.

According to Mayfield et al. (2012), users with high incoming stress tend to request less information from others, as a percentage of their time, and share much more information, in absolute terms. In addition, high information sharing has been shown to be a good predictor of stress reduction at the end of the chat (Mayfield et al., 2012). Regarding information sharing, we rely on Dialogue Acts (Austin, 1975) to model the speaker's intention in producing an utterance. In particular, we are interested in Dialogue Act label that is defined to represent descriptive, narrative, or personal information – the *statement*.

Dialogue Acts have been applied to the analysis of spoken (Stolcke et al., 2000; Cervone et al., 2018) as well as on-line written synchronous conversations (Forsythand and Martell, 2007). We apply Dialogue Act tag set defined in Forsythand and Martell (2007) to the analysis of our on-line asynchronous conversations. We argue that Dialogue Acts can be used to analyse user behaviour in social media and verify the presence of therapeutic factors.

3 Methodology

We select the three therapeutic factors – Universality, Altruism and Instillation of Hope – that can be best approximated using NLP techniques: Sentiment Analysis and Dialogue Act tagging. We discuss each one of the selected therapeutic factors and the identified necessary conditions. The listed conditions, however, are not sufficient to attribute the presence of a therapeutic factor with high confidence, which only can be obtained using expert annotation. Our analysis focuses on the structure of conversations; though content plays an important role as well.

Universality consists in the disconfirmation of patients' belief of uniqueness of their disease. This therapeutic factor is shown to be a powerful source of relief for the patient, according to Yalom and Leszcz (2005). From this definition, we can draw the following conditions that are applicable to our environment:

1. improvement of original poster's sentiment: we hypothesize that the discovery that other people passed through similar issues leads to a higher sentiment score;

2. posts containing negative personal experiences: to disconfirm the belief of uniqueness users have to share their story;

3. comments containing negative statements: to disconfirm the patient's feelings of uniqueness, the commenting user must tell a similar negative personal experience. This condition requires two sub-conditions: high presence of statements in comments and the presence of negative comments replying to negative posts.

Instillation of Hope is based on inspiration provided to participants by their peers. Through the inspiration provided by their peers, patients can increase their expectation on the therapy outcome. Yalom and Leszcz (2005) in several studies have demonstrated that a high expectation of help before the start of a therapy is significantly correlated with a positive therapy outcome. The author states that many patients pointed out the importance of having observed the improvement of others. Therefore, the three main conditions are the following:

1. improvement of original poster's sentiment: we hypothesize that instillation of hope leads to a higher sentiment score;

2. posts containing negative personal experiences: hope can be instilled in someone who shares a negative personal experience;

3. comments containing positive personal experiences: in order to instill hope, commenting posters must show to original posters an overall positive personal experience. To detect positive personal experience, we require the presence of statements in comments and a positive sentiment of comments replying to negative posts.

Altruism consists of peers offering support, reassurance, suggestions and insight, since they share similar problems with one another (Yalom and Leszcz, 2005). The experience of finding that a patient can be of value to others is refreshing and boosts self-esteem (Yalom and Leszcz, 2005). However, in the current study we focus on testing whether commenting posters are altruists or not. We do not test whether the altruistic behavior leads to an improvement on the altruist itself. For these reasons, we define three main conditions:

1. improvement of original poster's sentiment: we hypothesize that supportive and reassuring statements improve the sentiment score of the original poster;

2. posts contains negative personal experiences: users offer support, reassurance and suggestion when facing a negative personal experience of the original poster;

3. comments containing positive statements: either supportive or reassuring statements show by definition a positive intended emotional communication. Thus comments to the post should consist of positive sentiment statements.

Consequently, a conversation containing the aforementioned therapeutic factors should satisfy the following conditions in terms of NLP: Sentiment Analysis and Dialogue Acts.

1. original posters have a higher sentiment score at the end of the thread than at the beginning;

2. the original post consists mostly of polarised statements;

3. the presence of a significant amount of statements in comments, since both support and sharing similar negative experiences can be represented as statements;

4. both negative and positive statements in comments lead to higher final sentiment score of the original poster.

4 Datasets

We verify the presence of therapeutic factors in two social media datasets: OSG and Twitter. The first dataset is crawled from an on-line support groups website, and the second dataset consists of a small sample of Twitter conversation threads. Since the former consists of multi-threaded conversations, we apply a pre-processing to extract conversation threads to provide a fair comparison with the Twitter dataset. An example conversation from each data source is presented in Figure 1.

4.1 Twitter

We have downloaded 1,873 Twitter conversation threads, roughly 14k tweets, from a publicly available resource[3] that were previously pre-processed and have conversation threads extracted. A conversation in the dataset consists of at least 4 tweets. Even though, according to Paul and Dredze (2011), Twitter is broadly applicable to public health research, our expectation is that it contains less therapeutic conversations in comparison to specialized on-line support forums.

4.2 OSG

Our data has been developed by crawling and pre-processing an OSG web forum. The forum has a great variety of different groups such as depression, anxiety, stress, relationship, cancer, sexually transmitted diseases, etc. Each conversation starts with one post and can contain multiple comments. Each post or comment is represented by a poster, a timestamp, a list of users it is referencing to, thread id, a comment id and a conversation id. The thread id is the same for comments replying to each other, otherwise it is different. The thread id is increasing with time. Thus, it provides ordering

among threads; whereas the timestamp provides ordering in the thread.

Each conversation can belong to multiple groups. Consequently, the dataset needs to be processed to remove duplicates. The dataset resulting after de-duplication contains 295 thousand conversations, each conversation contains on average 6 comments. In total, there are 1.5 million comments. Since the created dataset is multi-threaded, we need to extract conversation threads, to eliminate paths not relevant to the original post.

4.2.1 Conversation Thread Extraction

The thread extraction algorithm is heuristic-based and consists of two steps: (1) creation of a tree, based on a post written by a user and the related comments and (2) transformation of the tree into a list of threads.

The tree creation is an extension of the approach of Gómez et al. (2008), where first a graph of conversation is constructed. In the approach, direct replies to a post are attached to the first nesting level and subsequent comments to increasing nesting levels. In our approach, we also exploit comments' features.

The tree creation is performed without processing the content of comments, which allows us to process posts and comments of any length efficiently. The heuristic used in the process is based on three simplifying assumptions:

1. Unless there is a specific reference to another comment or a user, comments are attached to the original post.

2. When replying, the commenting poster is always replying to the original post or some other comment. Unless specified otherwise, it is assumed that it is a response to the previous (in time) post/comment.

3. Subsequent comments by the same poster are part of the same thread.

To evaluate the performance of the thread extraction algorithm, 2 annotators have manually constructed the trees for 100 conversations. The performance of the algorithm on this set of 100 conversations is evaluated using accuracy and standard Information Retrieval evaluation metrics of precision, recall, and F_1 measure. The results are reported in Table 1 together with random and majority baselines. The turn-level percent agreement between the 2 annotators is 97.99% and Cohen's Kappa Coefficient is 83.80%.

---

[3] https://github.com/Phylliida/Dialogue-Datasets

SCENE – I OSG

ALICE : I want to tell him that if he can't have a real conversation with me then don't talk to me, because it hurts more to feel like I'm an obligation.... I don't want anyone to ever get close to me but I don't want to be alone.

BOB to feel like an obligation is really disheartening and takes a stab at the self esteem. What about the conversations make you feel like a an obligation? Have you talked to him about this?

ALICE He doesn't have a conversation, I know he doesn't mean to, he's just always busy now... I don't want to make him feel bad.

BOB Just remember that your needs matter too!

...

ALICE @Bob Thank you :)

SCENE – II TWITTER

CAROL : lol my best friend at the time got cheated on, we wrote on the guys truck..he started chasing us, i tripped and broke my ankle #justmyluck

DAVE wtf

CAROL it was ridiculous and i drove with my foot hanging out the window all f***ed up

DAVE when was this i'm so confused

Figure 1: Two example conversation threads extracted from an OSG and Twitter.

Approach	Acc	P	R	F_1
Majority Baseline	0.92	0.46	0.46	0.46
Random Baseline	0.87	0.14	0.14	0.14
Our Approach	**0.97**	**0.79**	**0.80**	**0.80**

Table 1: Performance of the thread extraction algorithm on a set of 100 manually constructed trees.

4.3 Data Representation

For both data sources, Twitter and OSG with extracted threads, posts and comments are tokenized[4] and sentence split. Each sentence is passed through Sentiment Analysis and Dialogue Act tagging. Since a post or a comment can contain multiple sentences, therefore multiple Dialogue Acts, it is represented as as a one-hot encoding, where each position represents a Dialogue Act.

For Sentiment Analysis we use a lexicon-based sentiment analyser introduced by Alistair and Diana (2005). For Dialogue Act tagging, on the other hand, we make use of a model trained on NPSChat corpus (Forsythand and Martell, 2007) following the approach of Lan et al. (2008).[5]

---

[4]NLTK sentence tokenizer

[5]The model achieves 80.21% accuracy.

5 Analysis

As we mentioned in Section 3, the presence of each of the therapeutic conditions under analysis is a necessary for a conversation to be considered to have therapeutic factors. In this section we present the results of our analysis with respect to these conditions.

5.1 Change in Sentiment score of Original Posters

The first condition which we test is the sentiment change in conversation threads, comparing the initial and final sentiment scores (i.e. posts' scores) of the original poster. The results of the analysis are presented in Figure 2. In the figure we can observe that the distribution of the sentiment change in the two datasets is different. While in Twitter the amount of conversations that lead to the increase of sentiment score is roughly equal to the amount of conversations that lead to the decrease of sentiment score; the situation is different for OSG. In OSG, the amount of conversations that lead to the increase of sentiment score is considerably higher.

Figure 3 provides a more fine grained analysis, where we additionally analyse the sentiment change in nominal polarity terms – negative

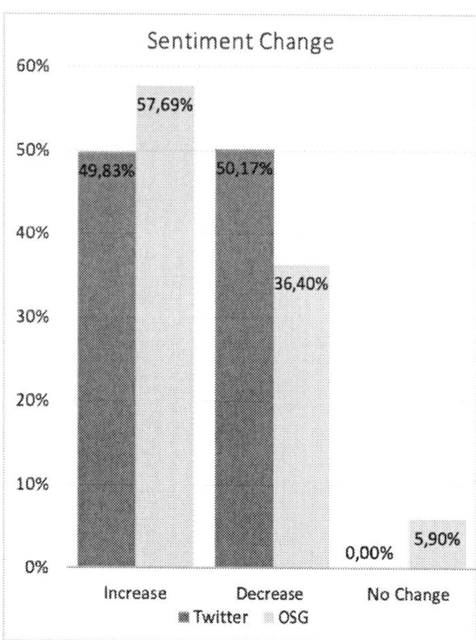

Figure 2: The percentages of threads in OSG and Twitter leading to the increase or decrease of the sentiment score of the original poster.

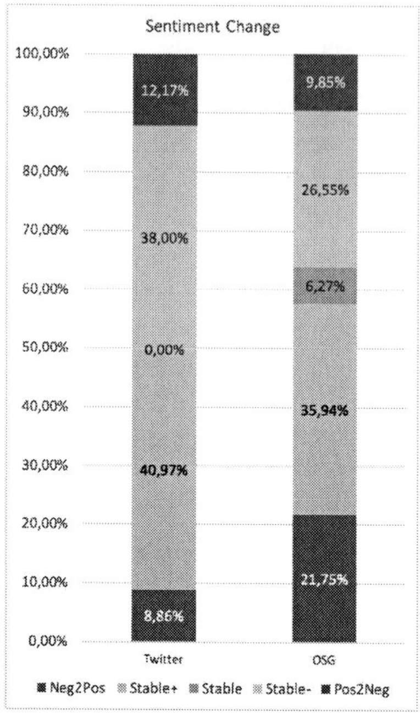

Figure 3: The sentiment polarity change in the two datasets - Twitter and OSG. Stable segments are labeled either as an increase (+), decrease (-) or no change in polarity, including neutral comments. Pos2Neg and Neg2Pos denote a nominal polarity change.

and positive. In OSG, the number of users that changed polarity from negative to positive is more than the double of the users that have changed the polarity from positive to negative. In Twitter, on the other hand, the users mostly changed polarity from positive to negative. Results of the analysis suggest that in OSG , sentiment increases and users tend to change polarity from negative to positive, whereas in Twitter sentiment tends to decrease. Verification of this condition alone indicates that the ratio of potentially therapeutic conversations in Twitter is lower.

5.2 Structure of Posts and Comments

Table 2 presents the distribution of automatically predicted per-sentence Dialogue Acts in the datasets. The most frequent tag is *statement* in both. In Table 3, on the other hand, we present the distribution of post and comment structures in terms of automatically predicted Dialogue Act tags. The structure is an unordered set of tags in the post or comment. From the table we can observe that the distribution of tag sets is similar between posts and comments. In both cases the most common set is *statement* only. However, conversations containing only *statement*, *emphasis* or *question* posts and comments predominantly appear in Twitter. Which is expected due to the

shorter length of Twitter posts and comments.

We can also observe that the original posters tend to ask more questions than the commenting posters – 19.83% for posts vs. 11.21% for comments (summed). This suggests that the original posters frequently ask either for suggestion or confirmation of their points of view or their disconfirmation. However, the high presence of personal experiences is supported by the high number of posts containing only statements.

High number of *statement* tags in comments suggests that users reply either with supporting or empathic statements or personal experience. However, 6.39% of comments contain *accept* and *reject* tags, which mark the degree to which a speaker accepts some previous proposal, plan, opinion, or statement (Stolcke et al., 2000). The described Dialogue Act tags are often used when commenting posters discuss original poster's point of view. For instance, "It's true. I felt the same." – {*Accept, Statement*} or "Well no. You're not alone" – {*Reject, Statement*}. The datasets differ with respect to the distribution of these Dialogue Acts tags, they appear more frequently in OSG.

Class	Twitter	OSG
Statement	62.9	73.0
Emphasis	9.6	6.3
ynQuestion	7.5	4.7
Continuer	2.5	4.3
whQuestion	6.1	3.7
Reject	2.6	2.9
Emotion	2.9	1.5
Accept	2.4	1.3
Greet	0.6	0.8
nAnswer	1.1	0.4
yAnswer	0.8	0.3
Bye	0.4	0.2
Clarify	0.1	< 0.1
Other	< 0.1	< 0.1

Table 2: The distribution (in percentages) of automatically predicted per-sentence Dialogue Act tags. Tags are counted separately for each sentence in the multi-sentence posts and comments.

5.3 Sentiment of Posts and Comments

Table 4 presents the distribution of sentiment polarity in post and comment statements (i.e. sentences tagged as *statement*). For OSG, the predominant sentiment label of statements is positive and it is the highest for both posts and comments. However, the difference between the amounts of positive and negative statements is higher for the replying comments (34.5% vs. 42.5%). For Twitter, on the other hand, the predominant sentiment label of statements is neutral and the polarity distribution between posts and comments is very close. One particular observation is that the ratio of negative statements is higher in OSG for both posts and comments than in Twitter, which supports the idea of sharing negative experiences.

Further we analyze whether the sentiment of a comment (i.e. the replying user) is affected by the sentiment of the original post (i.e. the user being replied to), which will imply that the users adapt their behaviour with respect to the post's sentiment. For the analysis, we split the datasets into three buckets according to the posts' sentiment score – negative, neutral, or positive, and represent each conversation in terms of percentages of comments (replies) with each sentiment label. The buckets are then compared using t-test for statistically significant differences.

Table 5 presents the distribution of sentiment labels with respect to the post's sentiment score.

Tag Set	Posts		Comments	
	Twitter	OSG	Twitter	OSG
Statement	64.12	38.79	57.14	41.45
Emphasis	3.01	1.31	4.42	3.96
ynQuestion	4.79	2.94	4.80	2.14
whQuestion	4.00	1.43	4.86	2.07
Statement +				
Emphasis	2.17	3.96	3.65	5.57
Continuer	0.99	6.29	0.92	4.59
ynQuestion	2.86	7.04	1.92	4.05
whQuestion	4.00	3.98	1.56	2.95
Accept	0.44	0.81	0.19	1.92
Reject	1.28	3.00	0.95	3.38

Table 3: The distribution (in percentages) of post and comment structures represented as unordered set of Dialogue Act tags.

	Sentiment Polarity		
	Negative	Neutral	Positive
OSG			
Posts	32.1	33.5	34.5
Comments	25.8	31.7	42.5
Twitter			
Posts	20.5	44.0	35.5
Comments	21.1	45.9	33.0

Table 4: The distribution (in percentages) of sentiment in *statement* sentences of posts and comments.

The patterns of distribution are similar across the datasets. We can observe that overall, replies tend to have a positive sentiment, which suggests that replying posters tend to have a positive attitude. However, the ratio of positive comments is higher for OSG than for Twitter.

The results of the Welch's t-test on OSG data reveal that there are statistically significant differences in the distribution of replying comments' sentiment between conversations with positive and negative starting posts. A positive post tends to get significantly more positive replies. Similarly, a negative post tends to get significantly more negative replies (both with $p < 0.01$).

Table 6 presents the distribution of the sentiment labels of the final text provided by the original poster with respect to the sentiment polarity of the comments. The results indicate that OSG participants are more supportive, as the majority of conversations end in a positive final sentiment regardless of the sentiment of comments. We can

Posts	Comments		
	Negative	Neutral	Positive
OSG			
Negative	27.25	14.87	57.88
Neutral	21.37	23.49	55.14
Positive	22.79	19.62	65.17
Twitter			
Negative	32.92	22.85	44.23
Neutral	26.48	25.00	48.52
Positive	18.79	16.04	57.60

Table 5: The distribution (in percentages) of reply sentiment labels with respect to the post's sentiment label.

Comments	Final Sentiment of OP		
	Negative	Neutral	Positive
OSG			
Negative	28.98	18.27	52.75
Neutral	25.20	25.70	49.10
Positive	22.60	18.01	59.38
Twitter			
Negative	24.14	41.78	34.08
Neutral	21.34	49.16	29.50
Positive	20.25	35.20	44.55

Table 6: The distribution (in percentages) of sentiment labels of the final text of the original poster (OP) with respect to the comment's sentiment label.

also observe that negative comments in OSG lead to positive sentiment, which supports the idea of sharing the negative experiences, thus presence of therapeutic factors. For Twitter, on the other hand, only positive comments lead to the positive final sentiments, whereas other comments lead predominantly to neutral final sentiments.

Our analysis in terms of sentiment and Dialogue Acts supports the presence of the three selected therapeutic factors – Universality, Altruism and Instillation of Hope – in OSG more than in Twitter. The main contributors to this conclusion are the facts that there is more positive change in the sentiment of the original posters in OSG (people seeking support) and that in OSG even negative and neutral comments are likely to lead to positive changes.

6 Conclusion

In this work, we propose a methodology to automatically analyse online social platforms for the presence of therapeutic factors (i.e. Universality, Altruism and Instillation of Hope). We evaluate our approach on two on-line platforms, Twitter and an OSG web forum. We apply NLP techniques of Sentiment Analysis and Dialogue Act tagging to automatically verify the presence of therapeutic factors, which allows us to analyse larger amounts of conversational data (as compared to previous studies).

Our analysis indicates that OSG conversations satisfy higher number of conditions approximating therapeutic factors than Twitter conversations. Given this outcome, we postulate that users who join support group websites spontaneously seem to benefit from it. Indeed, as shown in Section 5, the original posters who interact with others by replying to comments, have benefited from an improvement of their emotional state.

We would like to reemphasise that the conditions for the therapeutic factors are necessary but not sufficient; since our analysis focuses on the structure of conversations, being agnostic to the content. NLP, however, allows us to strengthen our approximations even further. Thus, the further extension of our work is also augmentation of our study with other language analysis metrics and their correlation with human annotation.

It should be noted that the proposed approach is an approximation of the tedious tasks of annotation of conversations by experts versed in the therapeutic factors and their associated theories. Even though we can use Sentiment Analysis to detect the existence of therapeutic factors, we cannot differentiate between Altruism and Instillation of Hope, as this requires differentiation between emotional state of the user and the intended emotional communication. Thus, the natural extensions of this work are differentiation between different therapeutic factors and comparison of the proposed analysis to the human evaluation.

Although we acknowledge that the proposed methodology does not serve as a replacement of manual analysis of OSG for the presence of therapeutic factors, we believe that it could facilitate and supplement this process. The method can serve as a tool for general practitioners and psychologists who can use it as an additional source of information regarding their patients condition and, in turn, offer a more personalised support that is better tailored to individual therapeutic needs.

References

Kennedy Alistair and Inkpen Diana. 2005. Sentiment classification of movie and product reviews using contextual valence shifters. *Proceedings of FINEXIN*.

Yair Amichai-Hamburger, Anat Brunstein Klomek, Doron Friedman, Oren Zuckerman, and Tal Shani-Sherman. 2014. The future of online therapy. *Computers in Human Behavior*, 41:288 – 294.

Gerard Andersson. 2017. Maximizing e-therapy. European Congress of psychology.

John Langshaw Austin. 1975. *How to do things with words*. Oxford university press.

Mitja D Back, Juliane M Stopfer, Simine Vazire, Sam Gaddis, Stefan C Schmukle, Boris Egloff, and Samuel D Gosling. 2010. Facebook profiles reflect actual personality, not self-idealization. *Psychological science*.

Azy Barak, Meyran Boniel-Nissim, and John Suler. 2008. Fostering empowerment in online support groups. *Computers in human behavior*, 24(5):1867–1883.

Alessandra Cervone, Enrico Gambi, Giuliano Tortoreto, Evgeny A Stepanov, and Giuseppe Riccardi. 2018. Automatically predicting user ratings for conversational systems. In *CLiC-it*.

Jae Eun Chung. 2013. Social networking in online support groups for health: How online social networking benefits patients. 19.

Micha Elsner and Eugene Charniak. 2010. Disentangling chat. *Comput. Linguist.*, 36(3):389–409.

Colin A Espie, Simon D Kyle, Chris Williams, Jason C Ong, Neil J Douglas, Peter Hames, and JS Brown. 2012. A randomized, placebo-controlled trial of online cognitive behavioral therapy for chronic insomnia disorder delivered via an automated media-rich web application. *Sleep*, 35(6):769–781.

Eric N Forsythand and Craig H Martell. 2007. Lexical and discourse analysis of online chat dialog. In *International Conference on Semantic Computing (ICSC 2007)*, pages 19–26. IEEE.

Nria Gala, Reinhard Rapp, and Gemma Bel-Enguix. 2014. *Language Production, Cognition, and the Lexicon*. Springer Publishing Company, Incorporated.

Juneau M Gary and Linda Remolino. 2000. Online support groups: Nuts and bolts, benefits, limitations and future directions. eric/cass digest.

Arindam Ghosh, Evgeny A Stepanov, Morena Danieli, and Giuseppe Riccardi. 2017. Are you stressed? detecting high stress from user diaries. In *2017 8th IEEE International Conference on Cognitive Infocommunications (CogInfoCom)*. IEEE.

Vicenç Gómez, Andreas Kaltenbrunner, and Vicente López. 2008. Statistical analysis of the social network and discussion threads in slashdot. In *Proceedings of the 17th International Conference on World Wide Web*, WWW '08, pages 645–654, New York, NY, USA. ACM.

David W Kissane, Brenda Grabsch, David M Clarke, Graeme C Smith, Anthony W Love, Sidney Bloch, Raymond D Snyder, and Yuelin Li. 2007. Supportive-expressive group therapy for women with metastatic breast cancer: survival and psychosocial outcome from a randomized controlled trial. *Psycho-Oncology*, 16(4):277–286.

Kwok Cheung Lan, Kei Shiu Ho, Pong Luk, Robert Wing, and Hong Va Leong. 2008. Dialogue act recognition using maximum entropy. *Journal of the American Society for Information Science and Technology*, 59(6):859–874.

Anderson Loos. 2003. *Glossary of linguistic terms*. SIL International.

Elijah Mayfield, Miaomiao Wen, Mitch Golant, and Carolyn Penstein Rosé. 2012. Discovering habits of effective online support group chatrooms. In *Proceedings of the 17th ACM International Conference on Supporting Group Work*, GROUP '12, pages 263–272, New York, NY, USA. ACM.

Wilson McDermut, Ivan W. Miller, and Richard A. Brown. 2001. The efficacy of group psychotherapy for depression: A meta-analysis and review of the empirical research. *Clinical Psychology: Science and Practice*, 8(1):98–116.

Praic McMahon. 2016. More young people seeking mental health support online. [Online; Retrieved on September 7, 2016].

Michael J Paul and Mark Dredze. 2011. You are what you tweet: Analyzing twitter for public health. In *Fifth International AAAI Conference on Weblogs and Social Media*.

Evgeny A Stepanov, Stephane Lathuiliere, Shammur Absar Chowdhury, Arindam Ghosh, Radu-Laurențiu Vieriu, Nicu Sebe, and Giuseppe Riccardi. 2018. Depression severity estimation from multiple modalities. In *2018 IEEE 20th International Conference on e-Health Networking, Applications and Services (Healthcom)*, pages 1–6. IEEE.

Andreas Stolcke, Noah Coccaro, Rebecca Bates, Paul Taylor, Carol Van Ess-Dykema, Klaus Ries, Elizabeth Shriberg, Daniel Jurafsky, Rachel Martin, and Marie Meteer. 2000. Dialogue act modeling for automatic tagging and recognition of conversational speech. *Comput. Linguist.*, 26(3):339–373.

M. Tartakovsky. 2016. 5 benefits of group therapy. [Online; Retrieved on September 5, 2016].

David L Vogel, Stephen R Wester, and Lisa M Larson. 2007. Avoidance of counseling: Psychological factors that inhibit seeking help. *Journal of Counseling & Development*, 85(4):410–422.

Irvin D Yalom and Molyn Leszcz. 2005. *The Theory and practice of Group Psychotherapy*. Basic books.

Transfer learning for health-related Twitter data

Anne Dirkson & Suzan Verberne
LIACS, Leiden University
Niels Bohrweg 1, Leiden, the Netherlands
{a.r.dirkson, s.verberne}@liacs.leidenuniv.nl

Abstract

Transfer learning is promising for many NLP applications, especially in tasks with limited labeled data. This paper describes the methods developed by team TMRLeiden for the 2019 Social Media Mining for Health Applications (SMM4H) Shared Task. Our methods use state-of-the-art transfer learning methods to classify, extract and normalise adverse drug effects (ADRs) and to classify personal health mentions from health-related tweets. The code and fine-tuned models are publicly available.[1]

1 Introduction

Transfer learning is promising for NLP applications, as it enables the use of universal pre-trained language models (LMs) for domains that suffer from a shortage of annotated data or resources, such as health-related social media. Universal LMs have recently achieved state-of-the-art results on a range of NLP tasks, such as classification (Howard and Ruder, 2018) and named entity recognition (NER) (Akbik et al., 2018). For the Shared Task of the 2019 Social Media Mining for Health Applications (SMM4H) workshop team TMRLeiden focused on employing state-of-the-art transfer learning from universal LMs to investigate its potential in this domain.

2 Task descriptions

ADR extraction The purpose of **Subtask 1** (S1) is to classify tweets as containing an adverse drug response (ADR) or not. Subsequently, these ADR mentions are extracted in **Subtask 2** (S2) and normalized to MedDRA concept IDs in **Subtask 3** (S3). MedDRA (Medical Dictionary for Regulatory Activities) is an international, standardized medical terminology.[2]

---

[1]https://github.com/AnneDirkson/
SharedTaskSMM4H2019
[2]https://www.meddra.org/

Personal Health Mention Extraction The goal of **Subtask 4** (S4) is to identify tweets that are personal health mentions, i.e. posts that mention a person who is affected as well as their specific condition (Karisani and Agichtein, 2018), as opposed to posts discussing health issues in general. Generalisability to both future data and different health domains is evaluated by including data from the same domain collected years after the training data, as well as data from entirely different disease domain.

3 Our approach

3.1 Preprocessing

We preprocessed all Twitter data using the lexical normalization pipeline by Sarker (2017). We also employed an in-house spelling correction method (Dirkson et al., 2019). Additionally, punctuation and non-UTF-8 characters were removed using regular expressions.

3.2 Additional Data

Personal Health Mentions For S4, the training data consists of data from one disease domain, namely influenza, in two contexts: having a flu infection and getting a flu vaccination. To improve generalisability, we supplemented this data with six labelled data sets from different disease domains (Karisani and Agichtein, 2018). We refer to this combined data set as S4+. For each subset, 10% was used for a combined validation set. For fine-tuning the ULMfit universal language model based on 28,595 Wikipedia articles (Wikitext-103) (Merity et al., 2017b), the DIEGO Drug Chatter corpus (Sarker and Gonzalez, 2017) was combined with the data from S1 and S4+ to form a larger unsupervised corpus of health-related Twitter data ('TwitterHealth'). For S4, fine-tuning was also attempted with only the S4+ data.

Proceedings of the 4th Social Media Mining for Health Applications (#SMM4H) Workshop & Shared Task, pages 89–92
Florence, Italy, August 2, 2019. ©2019 Association for Computational Linguistics

	S1	S2*	S3	S4	S4+
Dev	-	130	76	-	-
Train	14,634	910	1,756	6,996	11,832
Validation	1,626	130	76	777	1,314
Test	5000	1000	1000	TBA	TBA

Table 1: Data sets. *Only tweets containing ADRs were used for developing the system. TBA: To be announced

Concept Normalization The MedDRA concept names and their aliases in both MedDRA and the Consumer Health Vocabulary[3] were used to supplement the data from S3. This data set is hereafter called S3+.

3.3 Text Classification

Text classification was performed with fast.ai ULMfit (Howard and Ruder, 2018). As recommended, the initial learning rate (LR) of 0.01 was determined manually by inspecting the log LR compared to the loss. Default language models were fine-tuned using AWD_LSTM (Merity et al., 2017a) with (1) 1 cycle (LR = 0.01) for the last layer and then (2) 10 cycles (LR = 0.001) for all layers.

Subsequently, this model is used to train a classifier with F_1 as the metric, a dropout of 0.5 and a momentum of (0.8,0.7), in line with the recommendations. Training is done with (1) 1 cycle (LR = 0.02) on the last layer; (2) unfreezing of the second-to-last layer; (3) another cycle running from a 10-fold decrease of the previous LR to this LR divided by 2.6[4] (as recommended in the fast.ai MOOC).[4] This is repeated for the next layer and then for all layers. The last step consists of multiple cycles until F_1 starts to drop.

As an alternative classifier for S1, we used the absence of ADRs (noADE) according to the Bert embeddings NER method (see below) which was developed for the subsequent sub-task (S2) and aims to extract these ADR mentions. As a baseline for text classification, we used a Linear SVC with unigrams as features. The C parameter was tuned with a grid of 0.0001 to 1000 (steps of x10).

3.4 Named Entity Recognition

For S2, we experimented with different combinations of state-of-the-art Flair embeddings (Akbik et al., 2018), classical Glove embeddings and Bert embeddings (Devlin et al., 2018) using the Flair package. We used pre-trained Flair embeddings based on a mix of Web data, Wikipedia and subtitles; and the 'bert-base-uncased' variant of Bert embeddings. We also experimented with Flair embeddings combined with Glove embeddings (dimensionality of 100) based on FastText embeddings trained on Wikipedia (GloveWiki) or on Twitter data (GloveTwitter). Training for all embeddings was done with initial LR of 0.1, batch size of 32 and max epochs set to 150.

As a baseline for NER, we used a CRF with the default L-BFGS training algorithm with Elastic Net regularization. As features for the CRF, we used the lowercased word, its suffix, the word shape and its POS tag.[5]

3.5 Concept normalization

For S3, pre-trained Glove embeddings were used to train document embeddings on the extracted ADR entities in the S3 data including or excluding the aliases from CHV (S3+) with concept IDs as labels. We used the default RNN in Flair with a hidden size of 512. Glove embeddings (dim = 100) were based on FastText embeddings trained on Wikipedia. Token embeddings were re-projected (dim = 256) before inputting to the RNN.

4 Results

	Method	F_1 (range)	P	R
Average*		0.502	0.535	0.505
		(0.331)		
Run1	ULMfit[1]	**0.533**	**0.642**	0.455
Run2	noADE	0.418	0.284	**0.792**

Table 2: Results for ADR Classification (S1). *over all runs submitted [1]TwitterHealth data

For all four subtasks, our best transfer learning system consistently performs better than the average over all runs submitted to SMM4H. For classifying ADR mentions, our overall best performing system is a ULMfit model trained on the Twitter-Health corpus (see Table 2). Yet, the highest recall is attained by using the absence of named entities (noADE) as a classifier. This is in line with our validation results (see Table 6). For extracting ADRs, our best system is a combination of Bert with Flair embeddings without a separate classifier

---

[3]https://www.nlm.nih.gov/research/umls/sourcereleasedocs/current/CHV/

[4] https://course.fast.ai/

[5]https://sklearn-crfsuite.readthedocs.io/en/latest/tutorial.html

	Method	relaxed F_1 (range)	relaxed P	relaxed R	strict F_1 (range)	strict P	strict R
Average*		0.538 (0.486)	0.513	0.615	0.317 (0.422)	0.303	0.358
Run1	Bert+Flair[+]	**0.625**	0.555	**0.715**	**0.431**	0.381	**0.495**
Run2	Bert[+]	0.622	0.560	0.701	0.427	0.382	0.484
Run3	Bert+ADRClassifier	0.604	**0.718**	0.521	0.417	**0.494**	0.360

Table 3: Results for ADR Extraction(S2). *over all runs submitted [+]No separate classifier for sentences containing ADRs

	Method	relaxed F_1 (range)	relaxed P	relaxed R	strict F_1 (range)	strict P	strict R
Average*		0.297 (0.242)	0.291	0.312	0.212 (0.247)	0.205	0.224
Run1[+]	RNN Docembeddings	**0.312**	**0.370**	0.270	**0.250**	**0.296**	0.216
Run2[+]	RNN Docembeddings	0.303	0.272	0.343	0.244	0.218	0.277
Run3[+]	RNN Docembeddings	0.302	0.267	**0.347**	0.246	0.218	**0.283**

Table 4: Results for concept normalization (S3). *over all runs submitted [+]Runs same as S2 prior to concept normalization

	Method		Acc. (range)	F_1 (range)	P	R
Average*			0.781 (0.263)	0.701 (0.464)	0.902	0.585
Run1	ULMfit with S4+ data	Domain1	0.869	0.859	0.952	0.781
		Domain2	0.638	0.419	0.750	0.290
		Domain3	0.786	0.539	1.000	0.368
		Mean	**0.793**	**0.726**	**0.940**	**0.591**
Run2	ULMfit with TwitterHealth data	Domain1	0.863	0.849	0.969	0.756
		Domain2	0.609	0.342	0.700	0.226
		Domain3	0.768	0.480	1.000	0.316
		Mean	0.786	0.716	0.928	0.583

Table 5: Results for personal health mention classification (S4). *over all runs submitted

for sentences containing ADR mentions (see Table 3). However, using Bert embeddings alone *with* the ULMfit classifier from S1 appears to be more precise. During validation, we found that combinations of Glove embeddings (based on Twitter or Wikipedia) and Flair embeddings performed poorly compared to the submitted systems (see Table 7). For mapping the ADRs to MedDRA concepts, we only submitted one system with different preceding NER models (see Table 4), since adding the alias information (S3+) decreased both precision and recall (see Table 8). Our RNN document embeddings with only the S3 data, however, performed better than average. Lastly, for the classification of personal health mentions, our best classifier was a ULMfit model fine-tuned on the S4+ data (see Table 5), which outperformed the average result and the ULMfit model trained on the larger TwitterHealth corpus on all metrics. This system similarly outperformed the other ULMfit model on the validation data (see Table 9).

Method	F_1	P	R
Baseline: Linear SVC (C=1.0)	0.475	0.526	0.433
ULMfit[1]	**0.574**	**0.574**	0.574
noADE	0.330	0.207	**0.823**

Table 6: Validation results for ADR classification (S1)
[1]TwitterHealth data

Method	Micro-F_1	P	R
Baseline: CRF	0.235	0.560	0.149
Flair+ GloveWiki	0.596	0.666	0.540
Flair+ GloveTwitter	0.577	0.655	0.515
Bert	0.640	**0.699**	0.590
Bert+Flair	**0.649**	**0.699**	**0.606**

Table 7: Validation results for ADR extraction (S2)

Method	F_1	P	R
RNNDocembeddings with S3	**0.623**	**0.566**	**0.694**
RNNDocembeddings with S3+	0.253	0.171	0.482

Table 8: Validation results for concept normalization (S3)

Method	F_1	P	R
Baseline: Linear SVC (C=0.1)	0.615	0.678	0.572
ULMfit with S4+ data	**0.712**	**0.743**	**0.701**
ULMfit with TwitterHealth data	0.692	0.738	0.676

Table 9: Mean validation results for personal health mention classification (S4) averaged over eight data sets of S4+

5 Conclusions

Transfer learning using default and recommended settings offers above average results for various NLP tasks using health-related Twitter data. More research is necessary to investigate whether state-of-the-art performance may be possible with further domain-specific adaptation, for instance by tuning hyper-parameters, training embeddings on medical data or by dealing with domain-specific vocabulary absent in the language model.

References

Alan Akbik, Duncan Blythe, and Roland Vollgraf. 2018. Contextual string embeddings for sequence labeling. In *Proceedings of the 27th International Conference on Computational Linguistics*, pages 1638–1649, Santa Fe, New Mexico, USA. Association for Computational Linguistics.

Jacob Devlin, Ming-Wei Chang, Kenton Lee, and Kristina Toutanova. 2018. BERT: Pre-training of Deep Bidirectional Transformers for Language Understanding. *ArXiv*.

Anne Dirkson, Suzan Verberne, Gerard van Oortmerssen, Hans van Gelderblom, and Wessel Kraaij. 2019. Lexical Normalization of User-Generated Medical Forum Data. In *Proceedings of 2019 ACL workshop Social Media Mining 4 Health (SMM4H)*.

Jeremy Howard and Sebastian Ruder. 2018. Universal language model fine-tuning for text classification. In *Proceedings of the 56th Annual Meeting of the Association for Computational Linguistics (Volume 1: Long Papers)*, pages 328–339, Melbourne, Australia. Association for Computational Linguistics.

Payam Karisani and Eugene Agichtein. 2018. Did you really just have a heart attack?: Towards robust detection of personal health mentions in social media. In *Proceedings of the 2018 World Wide Web Conference*, WWW '18, pages 137–146, Republic and Canton of Geneva, Switzerland. International World Wide Web Conferences Steering Committee.

Stephen Merity, Nitish Shirish Keskar, and Richard Socher. 2017a. Regularizing and Optimizing LSTM Language Models. *ArXiv*.

Stephen Merity, Caiming Xiong, James Bradbury, and Richard Socher. 2017b. Pointer Sentinel Mixture Models. In *ICLR 2017*.

Abeed Sarker. 2017. A customizable pipeline for social media text normalization. *Social Network Analysis and Mining*, 7(1):45.

Abeed Sarker and Graciela Gonzalez. 2017. A corpus for mining drug-related knowledge from Twitter chatter: Language models and their utilities. *Data in Brief*, 10:122–131.

NLP@UNED at SMM4H 2019: Neural Networks Applied to Automatic Classifications of Adverse Effects Mentions in Tweets

Javier Cortes-Tejada
NLP & IR Group
UNED
28040 Madrid, Spain
jcortes@lsi.uned.es

Juan Martinez-Romo
NLP & IR Group (UNED)
IMIENS
28040 Madrid, Spain
juaner@lsi.uned.es

Lourdes Araujo
NLP & IR Group (UNED)
IMIENS
28040 Madrid, Spain
lurdes@lsi.uned.es

Abstract

This paper describes a system for automatically classifying adverse effects mentions in tweets developed for the task 1 at Social Media Mining for Health Applications (SMM4H) Shared Task 2019. We have developed a system based on LSTM neural networks inspired by the excellent results obtained by deep learning classifiers in the last edition of this task. The network is trained along with Twitter GloVe pre-trained word embeddings.

1 Introduction

The Shared Task (Weissenbacher et al., 2019) of the 2019 Social Media Mining for Health Applications (SMM4H) Workshop proposed several Natural Language Processing (NLP) tasks using social media mining for health monitoring. Since these tasks involve NLP techniques, they are as interesting as difficult to solve because these systems should be able to work with many linguistics variations and model the different ways people express medical-related concepts in social media. In addition, we must take into account the level of noise caused by creative sentences, misspellings or ambiguous and sarcastic expressions which makes hard to tackle these tasks.

For this shared task we decided to participate in the first task. This task proposes to find tweets mentioning Adverse Drug Reactions (ADR), taking into account the linguistic variations between ADRs and indications (the reason to use the medication). We have developed a system based on LSTM networks due to their latest achievements in the last edition of this task (Xherija, 2018).

2 Dataset

In this section we describe the dataset of the task 1 and the applied pre-processing. This task proposes to find tweets mentioning ADRs, therefore we have to deal with raw text extracted from Twitter.

The publicly available dataset contains for each tweet: (i) the user ID, (ii) the tweet ID, and (iii) the binary annotation indicating the presence or absence of ADRs. The dataset contains 24606 tweets manually tagged, being around 10% (2358) of tweets mentioning ADRs, and around the remaining 90% (22248) are tweets without ADRs.

2.1 Pre-processing

Regarding the dataset we normalized typical Twitter strings such as @user by <USER>, #hashtag by <HASHTAG> or https://... by <URL> to decrease the vocabulary size and reduce the dataset variability by grouping several tokens under the same meaning.

We also handle several elongated words such as "my gooooood". In these cases we replaced each token by a unique representation, for example "aaargh" and "arrggggh" by "argh".

Finally the last step was to replace several constructions like "it's" by "it is" or "OMG" by "Oh my god" and tokenize the text. For this step we used regular expressions and *NLTK* (Loper and Bird, 2002) to tokenize the text. We used specifically the class *TweetTokenizer* which is especially useful processing tweets since it splits the text into tokens, as others tokenizers, but also it takes into account some text elements like emojis or exclamatory particles, which are correctly separated into new tokens.

We didn't remove any stop-word or convert to lowercase the text because that might change the meaning of a tweet drastically.

3 System architecture

We used a model based on a Bi-LSTM network due to its high performance in NLP tasks being

Proceedings of the 4th Social Media Mining for Health Applications (#SMM4H) Workshop & Shared Task, pages 93–95
Florence, Italy, August 2, 2019. ©2019 Association for Computational Linguistics

used along with Twitter GloVe (Pennington et al., 2014) embeddings. The input of the system is a tweet (a sequence of words) which is used by the Embedding Layer with a fixed input size, while the weights of this layer are given by the GloVe word embeddings trained with 2 billion tweets. We have chosen these embeddings instead of others like word2vec (Mikolov et al., 2013), godin (Godin et al., 2015) or shin (Shin et al., 2016) because Twitter GloVe is trained with tweets, what is very useful since it allows us to have a greater vocabulary and also more similar to the text provided by the task.

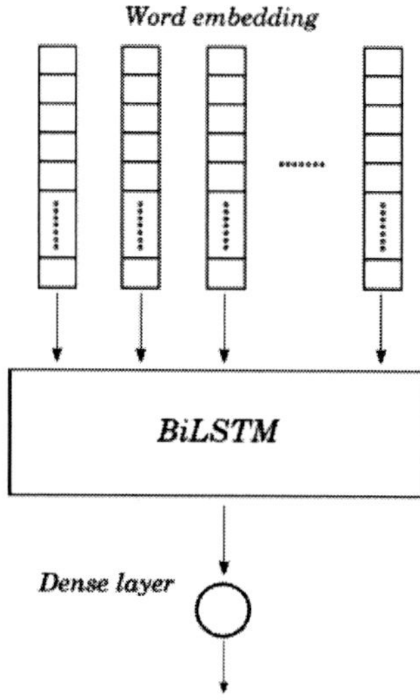

Figure 1: System architecture based on Twitter GloVe embeddings and a Bi-LSTM network.

As it can be seen in Figure 1, the next layer of our system is a Bi-LSTM layer. We decided to use it because a single LSTM network have not access to further tokens as they have not been seen. A Bi-LSTM has access to past tokens and future tokens, so this layer will give us a complete knowledge about the tweet; one LSTM will scan the sentence in one direction and the other will scan in the reverse direction. After these two layers we set a Dropout layer to prevent overfitting (Peng et al., 2015) with a rate of 0.3 for the Embedding layer and 0.5 for the Bi-LSTM layer. Finally we added a Dense layer with a sigmoid activation function at the end of the network to get the final results.

Regarding hyper parameters we used some configurations before we submitted the runs. For these tests we have tuned the epochs, the size of the batch (32, 64 and 128), the size of the embedding (vector of 50 and 100 dimensions in both embed-dings), and the optimizer by considering a couple of them as Adam (Kingma and Ba, 2014) and Ada-Grad (Duchi et al., 2011). We also handle the vocabulary tokens by adding pad right. At the end we chose the 3 configurations that reported the best results, whose hyper parameters are shown in Table 1.

4 Experiments and Results

For the implementation of the system we chose *Keras* and *Tensorflow* (Abadi et al., 2016) while for the pre-processing of the data we used *Scikit-learn* (Pedregosa et al., 2011), in particular for padding and split the dataset into validation, train and test sets.

In order to test the functioning of our system we used the evaluation script provided by the organizers. Several experiments are shown in Table 2. In these experiments we used a network without embeddings (Base) and with two types of embeddings, one pre-trained on Wikipedia pages (Wikipedia GloVe) and the other one based on tweets (Twitter GloVe). Due to the better performance shown by the configuration that used Twitter GloVe pre-trained embeddings, we decided to use it for the runs that we submitted to the task.

Table 3 shows the official results for the three runs that we submitted to the task 1 and the task average score provided by the organizers. According to the results obtained, it could be said that a greater number of epochs provides better results although the recall begins to fall.

5 Conclusions

Taking into account the experiments carried out on the training set and the results obtained, we can say that the use of embeddings pre-trained on tweets has been positive, that a greater number of epochs has provide us a better performance and that the best feature of our system is the recall as it obtains a value above the average.

In the future, we will try to create a more complex system to improve its performance. For this task we will add new features such as POS tagging and char embeddings as well as an attention mechanism.

Agreements

This work has been partially supported by the projects PROSA-MED (TIN2016-77820-C3-2-R) and EXTRAE (IMIENS 2017).

	Run 1	**Run 2**	**Run 3**
Epochs	40	30	20
Embedding	Twitter Glove		
Batch size	64	64	32
Embedding size	100	100	50
Optimizer	AdaGrad	AdaGrad	Adam

Table 1: Hyper parameter tunning used in the 3 runs submitted for task 1.

System	**P**	**R**	**F1**
Base	0.408	0.430	0.419
Base + Wikipedia G	0.450	0.512	0.483
Base + Twitter G	**0.458**	**0.590**	**0.510**

Table 2: System results according the Precision (P), Recall (R) and F-Measure (F1) scores.

References

Martín Abadi, Paul Barham, Jianmin Chen, Zhifeng Chen, Andy Davis, Jeffrey Dean, Matthieu Devin, Sanjay Ghemawat, Geoffrey Irving, Michael Isard, et al. 2016. Tensorflow: A system for large-scale machine learning. In *12th {USENIX} Symposium on Operating Systems Design and Implementation ({OSDI} 16)*, pages 265–283.

John Duchi, Elad Hazan, and Yoram Singer. 2011. Adaptive subgradient methods for online learning and stochastic optimization. *Journal of Machine Learning Research*, 12(Jul):2121–2159.

Fréderic Godin, Baptist Vandersmissen, Wesley De Neve, and Rik Van de Walle. 2015. Multimedia lab @ acl wnut ner shared task: Named entity recognition for twitter microposts using distributed word representations. In *Proceedings of the Workshop on Noisy User-generated Text*, pages 146–153.

Diederik P Kingma and Jimmy Ba. 2014. Adam: A method for stochastic optimization. *arXiv preprint arXiv:1412.6980*.

Edward Loper and Steven Bird. 2002. Nltk: the natural language toolkit. *arXiv preprint cs/0205028*.

Tomas Mikolov, Kai Chen, Greg Corrado, and Jeffrey Dean. 2013. Efficient estimation of word representations in vector space. *arXiv preprint arXiv:1301.3781*.

Fabian Pedregosa, Gaël Varoquaux, Alexandre Gramfort, Vincent Michel, Bertrand Thirion, Olivier Grisel, Mathieu Blondel, Peter Prettenhofer, Ron Weiss, Vincent Dubourg, et al. 2011. Scikit-learn: Machine learning in python. *Journal of machine learning research*, 12(Oct):2825–2830.

Hao Peng, Lili Mou, Ge Li, Yunchuan Chen, Yangyang Lu, and Zhi Jin. 2015. A comparative study on regularization strategies for embedding-based neural networks. *arXiv preprint arXiv:1508.03721*.

Jeffrey Pennington, Richard Socher, and Christopher Manning. 2014. Glove: Global vectors for word representation. In *Proceedings of the 2014 conference on empirical methods in natural language processing (EMNLP)*, pages 1532–1543.

Bonggun Shin, Timothy Lee, and Jinho D Choi. 2016. Lexicon integrated cnn models with attention for sentiment analysis. *arXiv preprint arXiv:1610.06272*.

Davy Weissenbacher, Abeed Sarker, Arjun Magge, Ashlynn Daughton, Karen O'Connor, Michael Paul, and Graciela Gonzalez-Hernandez. 2019. Overview of the fourth Social Media Mining for Health (SMM4H) shared task at ACL 2019. In *Proceedings of the 2019 ACL Workshop SMM4H: The 4th Social Media Mining for Health Applications Workshop Shared Task*.

Orest Xherija. 2018. Classification of medication-related tweets using stacked bidirectional lstms with context-aware attention. In *Proceedings of the 2018 EMNLP Workshop SMM4H: The 3rd Social Media Mining for Health Applications Workshop and Shared Task*, pages 38–42. Association for Computational Linguistics.

Runs	**P**	**R**	**F1**
Run 1 (30 epochs)	0.463	**0.535**	0.408
Run 2 (40 epochs)	**0.472**	0.524	**0.429**
Run 3 (20 epochs)	0.431	0.491	0.385
Task average score	0.535	0.505	0.501

Table 3: Official results for the three runs that participated in task 1 and task average score provided by organizers.

Detecting and Extracting of Adverse Drug Reaction Mentioning Tweets with Multi-Head Self-Attention

Suyu Ge[†], Tao Qi[†], Chuhan Wu, Yongfeng Huang
Department of Electronic Engineering, Tsinghua University Beijing 100084, China
{gesy17,qit16,wuch15,yfhuang}@mails.tsinghua.edu.cn

Abstract

This paper describes our system for the first and second shared tasks of the fourth Social Media Mining for Health Applications (SMM4H) workshop. We enhance tweet representation with a language model and distinguish the importance of different words with Multi-Head Self-Attention. In addition, transfer learning is exploited to make up for the data shortage. Our system achieved competitive results on both tasks with an F1-score of 0.5718 for task 1 and 0.653 (overlap) / 0.357 (strict) for task 2.

1 Introduction

Automatic adverse drug reaction (ADR) detection and extraction are of great social benefits to public health, with which pharmacovigilance (Sarker and Gonzalez, 2015) can be performed at a broader and more automatic level. Recent research focus their attention on online public sources such as tweets due to their availability and authenticity (Onishi et al., 2018; Adrover et al., 2015; Salathé and Khandelwal, 2011).

The SMM4H shared task is proposed (Weissenbacher et al., 2019) to enhance ADR recognition. Task 1 is a binary classification task between ADR mentioned tweets and drug name only tweets, followed by task 2 to extract the particular position of ADR entities. Based on the work we did last year (Wu et al., 2018), we extend our previous model with hierarchical tweet representation and multi-head self-attention (HTR-MSA) to a model using both hierarchical tweet representation and attention (HTA) to jointly participate both tasks. Moreover, additional features and a language model are incorporated to enhance the semantic representations. In task 1, transfer learning

on a smaller dataset is exploited. In task 2, we add a CRF layer for the named entity recognition task.

2 Our Approach

Our HTA model can be divided into the following three parts: hierarchical word representation, hierarchical tweet representation and tweet classification, which are introduced as follows.

2.1 Hierarchical Word Representation

In order to combat out-of-vocabulary medical terminology, misspellings and user created abbreviations, we propose a character modeling at a lower level before traditional word representation. We denote the character sequence of i_{th} word as $\mathbf{w}_i = [\mathbf{C}_{i,1}, \mathbf{C}_{i,2}, ..., \mathbf{C}_{i,N}]$, where N is the word length. A character embedding matrix $\mathbf{M}^c \in \mathcal{R}^{V \times D}$ is utilized to convert $\mathbf{w}_i$ into vector sequence $\mathbf{E}_i^c = [\mathbf{e}_{i,1}, \mathbf{e}_{i,2}, ..., \mathbf{e}_{i,N}]$, where V denotes the character vocabulary size and D denotes the dimension of character embedding.

After a character embedding is generalized, character-level convolutional neural network is employed to capture local combined character feature. Assuming the window size of CNN filters is $2w + 1$ and $\mathbf{U}_c, b_c$ are kernel and bias parameters respectively, a convolutional representation $\mathbf{h}_{i,j}$ of character embedding vectors from position $j - w$ to $j + w$ is formed as follows:

$$\mathbf{h}_{i,j} = ReLU(\mathbf{U}_c \times \mathbf{e}_{i,(j-w):(j+w)} + b_c) \quad (1)$$

To remove unnecessary information, we apply the max pooling to pertain only the most salient feature of the i_{th} word.

Other features are added at a word level, such as word2vec-twitter (Godin et al., 2015) word embedding, pos-tag from NLTK library (Bird et al., 2009) and sentiment lexicon[1]. To strengthen the

---

[†]Equal contribution.

[1]http://sentiwordnet.isti.cnr.it/

Proceedings of the 4th Social Media Mining for Health Applications (#SMM4H) Workshop & Shared Task, pages 96–98
Florence, Italy, August 2, 2019. ©2019 Association for Computational Linguistics

medical meaning of word representation, word appearance in SIDER 4.1 medical lexicon[2] is transformed to one-hot vector as additional feature. Besides, the language model ELMo embedding (Peters et al., 2018) is incorporated to overcome the shortage of limited data and get better semantic meaning. Since ELMo contains character level information in their model, it fits better to our task goal than other language model that utilizes a fixed word look-up dictionary.

The final output of our hierarchical word representation is the concatenation of character representation, word embedding, pos-tag, sentiment lexicon, medical lexicon feature and language model output.

2.2 Hierarchical Tweet Representation

We first send word representation obtained in the previous module to a Bi-LSTM layer to encode long-distance information. The Bi-LSTM output of a sentence of length M is denoted as $\mathbf{H} = [\mathbf{h}_1, \mathbf{h}_2, ..., \mathbf{h}_M]$.

The second layer takes advantage of multi-head self-attention (Vaswani et al., 2017) to mine internal relation between words in the same sentence. In our layout, the representation vector $\mathbf{m}_{i,j}$ of the j_{th} word learned by the i_{th} attention head is computed by weighted summation of $\mathbf{H}$:

$$\hat{\alpha}_{j,k}^i = \mathbf{h_j}^T \mathbf{U_i} \mathbf{h_k}, \tag{2}$$

$$\alpha_{j,k}^i = \frac{\exp(\hat{\alpha}_{j,k}^i)}{\Sigma_{m=1}^M \exp(\hat{\alpha}_{j,m}^i)}, \tag{3}$$

$$\mathbf{m}_{i,j} = \mathbf{W}_i(\Sigma_{m=1}^M \alpha_{j,m}^i \mathbf{h}_m), \tag{4}$$

$\mathbf{U}_i$ and $\mathbf{W}_i$ are the parameters of the i_{th} self-attention head, and $\alpha_{j,k}^i$ represents the related weight between j_{th} and k_{th} words. After concatenating outputs from h different self-attention heads, we get the representation $\mathbf{m}_j = [\mathbf{m}_{1,j}; \mathbf{m}_{2,j}; ...; \mathbf{m}_{h,j}]$ of the j_{th} word.

2.3 Tweet Classification

For task 1, we use an additive attention mechanism to selectively combine word representations. The model is trained with a cost-sensitive weighted loss function (Santos-Rodrguez et al., 2009). Sentence level binary labels are then generated for task 1. However, in task 2 word level labels are needed, so we use a CRF layer to predict word level entity tags after self-attention vectors produced in the lower level.

[2]http://sideeffects.embl.de/

3 Experiments

3.1 Experiment Settings

In our experiments,the word embedding we use is 400 dimension and Bi-LSTM network has 2×200 units. The CNN network has 400 filters with window size of 3. There are 16 heads in the multi-head self-attention network, and the output dimension of each head is 16. Adam is selected as the optimizer.

Transfer learning is conducted on the CADEC medical ADR dataset (Karimi et al., 2015) first in task 1. However, we do not adopt this method in task 2 due to the relative small training dataset of this task. For the word classification, we train for this task a marginal CRF with probabilities as output.

3.2 Experiment Results

Detailed evaluation score is illustrated in table 1, which illustrated the effectiveness of our approach. In task 1, our model outperforms the average score among all participants by 0.070. In task 2, the improvement on relax F1 is also significant, we improve 0.115 on relax F1 and 0.040 on strict F1. Besides, compared to the best model we submitted for task 1 last year (Wu et al., 2018), which reached a 0.522 F1 score, our method with the language model and transfer learning improves the original model by 0.050.

4 Conclusion

We design HTA, a hierarchical tweet representation and attention model for SMM4H shared task 1 and 2, our model attains high evaluation scores on both tasks and generates promising application value.

Acknowledgments

This work was supported by the National Key Research and Development Program of China under Grant number 2018YFC1604002, and the National Natural Science Foundation of China under Grant numbers U1836204, U1705261, U1636113, U1536201, and U1536207.

References

Cosme Adrover, Todd Bodnar, Zhuojie Huang, Amalio Telenti, and Marcel Salathé. 2015. Identifying adverse effects of hiv drug treatment and associated

	Task 1	**Task 2** (relax)	**Task 2** (strict)
Precision	0.467	0.612	0.329
Recall	0.738	0.698	0.390
F1 Score	0.572	0.653	0.357
Average F1 (mean)	0.502	0.538	0.317
F1 Range (mean)	0.3308	0.486	0.422

Table 1: Evaluation Results.

sentiments using twitter. *JMIR public health and surveillance*, 1(2):e7.

Steven Bird, Ewan Klein, and Edward Loper. 2009. *Natural language processing with Python: analyzing text with the natural language toolkit.* " O'Reilly Media, Inc.".

Frderic Godin, Baptist Vandersmissen, Wesley De Neve, and Rik Van De Walle. 2015. Multimedia lab @ acl w-nut ner shared task: Named entity recognition for twitter microposts using distributed word representations. In *Workshop on User-generated Text*.

Sarvnaz Karimi, Alejandro Metke-Jimenez, Madonna Kemp, and Chen Wang. 2015. Cadec: A corpus of adverse drug event annotations. *Journal of biomedical informatics*, 55:73–81.

Takeshi Onishi, Davy Weissenbacher, Ari Klein, Karen O'Connor, and Graciela Gonzalez-Hernandez. 2018. Dealing with medication non-adherence expressions in twitter. In *Proceedings of the 2018 EMNLP Workshop SMM4H: The 3rd Social Media Mining for Health Applications Workshop and Shared Task*, pages 32–33, Brussels, Belgium. Association for Computational Linguistics.

Matthew E. Peters, Mark Neumann, Mohit Iyyer, Matt Gardner, Christopher Clark, Kenton Lee, and Luke Zettlemoyer. 2018. Deep contextualized word representations. In *Proc. of NAACL*.

Marcel Salathé and Shashank Khandelwal. 2011. Assessing vaccination sentiments with online social media: implications for infectious disease dynamics and control. *PLoS computational biology*, 7(10):e1002199.

Ral Santos-Rodrguez, Dario Garca-Garca, and Jess Cid-Sueiro. 2009. Cost-sensitive classification based on bregman divergences for medical diagnosis.

Abeed Sarker and Graciela Gonzalez. 2015. Portable automatic text classification for adverse drug reaction detection via multi-corpus training. *Journal of biomedical informatics*, 53:196–207.

Ashish Vaswani, Noam Shazeer, Niki Parmar, Jakob Uszkoreit, Llion Jones, Aidan N Gomez, Łukasz Kaiser, and Illia Polosukhin. 2017. Attention is all you need. In *Advances in neural information processing systems*, pages 5998–6008.

Davy Weissenbacher, Abeed Sarker, Arjun Magge, Ashlynn Daughton, Karen O'Connor, Michael Paul, and Graciela Gonzalez-Hernandez. 2019. Overview of the fourth Social Media Mining for Health (SMM4H) shared task at ACL 2019. In *Proceedings of the 2019 ACL Workshop SMM4H: The 4th Social Media Mining for Health Applications Workshop and Shared Task*.

Chuhan Wu, Fangzhao Wu, Junxin Liu, Sixing Wu, Yongfeng Huang, and Xing Xie. 2018. Detecting tweets mentioning drug name and adverse drug reaction with hierarchical tweet representation and multi-head self-attention. In *Proceedings of the 2018 EMNLP Workshop SMM4H: The 3rd Social Media Mining for Health Applications Workshop & Shared Task*, pages 34–37.

Deep Learning for Identification of Adverse Effect Mentions in Twitter Data

Paul Barry and Ozlem Uzuner, PhD, FACMI
George Mason University
pbarry2, ouzuner@gmu.edu

Abstract

Social Media Mining for Health Applications (SMM4H) Adverse Effect Mentions Shared Task challenges participants to accurately identify spans of text within a tweet that correspond to Adverse Effects (AEs) resulting from medication usage (Weissenbacher et al., 2019). This task features a training data set of 2,367 tweets, in addition to a 1,000 tweet evaluation data set. The solution presented here features a bidirectional Long Short-term Memory Network (bi-LSTM) for the generation of character-level embeddings. It uses a second bi-LSTM trained on both character and token level embeddings to feed a Conditional Random Field (CRF) which provides the final classification. This paper further discusses the deep learning algorithms used in our solution.

1 Data

The training data consists of 2,367 unique tweets of which 1,212 are positive examples and 1,155 are negative while the evaluation data consists of 1,000 tweets with 500 positive examples and 500 negatives. Of the 1,212 positive examples in the training set, 345 examples present two or more spans within the tweet that are AEs experienced by the individual. The remaining positive examples contain only one AE span. Spans of AEs are not limited to singular words nor are they required to be whitespace delimited. Because of this, many AEs within the data set consist of multiple words. Spans are not limited to English words or whole words, so abbreviations, portions of words, and concatenations of multiple words are expected. Tweets provided to participants had all alphabetical characters converted to their lowercase form. No other preprocessing steps were performed prior to dataset distribution. We divided the training dataset into subsets with 1,657 tweets used for training,

355 for validation, and 355 for testing. We tuned our parameters on the training set and report final results on the shared task evaluation set.

2 Preprocessing

We preprocessed AEs to consolidate overlapping spans and remove AEs that are a subset of others. Subsequently, we replaced twitter handles with "@person" to reduce the noise inherent to multiple tokens sharing the same meaning and to reduce dimensionality. To further reduce dimensionality, we removed the URLs within tweets as they do not provide contextual value. The hashtag character, "#", was removed so hashtag words could be treated like regular words rather than as separate, unique tokens. Tokenization was performed and tested using several tokenizers to include the Natural Language Toolkit's (NTLK) Word Tokenizer, NLTK's Word Punct Tokenizer, NLTK's Whitespace Tokenizer, and the Stanford Tokenizer (Manning et al., 2014; Bird et al., 2009). Lastly, the removal of all special characters was evaluated in conjunction with each of the above methods.

3 System Structure

The system used in this study is based around a Recurrent Neural Network (RNN) variant known as a bi-LSTM which features the Long Short-term Memory (LSTM) unit (Hochreiter and Schmidhuber, 1997). The system consists of four layers: a character embedding layer, a token embedding layer, a label prediction layer, and a label sequence optimization layer (Dernoncourt et al., 2017). As input, it uses three portions of the dataset for training, validation, and testing. Input to the bi-LSTM consists of word embeddings. We initialized word embeddings based on pretrained GloVe embeddings (Pennington et al., 2014). We then used ELMo to continue training embeddings

Proceedings of the 4ᵗʰ Social Media Mining for Health Applications (#SMM4H) Workshop & Shared Task, pages 99–101
Florence, Italy, August 2, 2019. ©2019 Association for Computational Linguistics

so they better represent each word's usage within the corpus (Peters et al., 2018). The trained word embeddings are augmented by training a bi-LSTM model on individual characters within a word and concatenating the character embeddings onto the word embedding vector. These character-enhanced token-embeddings are then passed as input into a second bi-LSTM layer in which both directions predict the label. The output from both directions is concatenated and passed to a CRF which provides the model's final prediction (Dernoncourt et al., 2017).

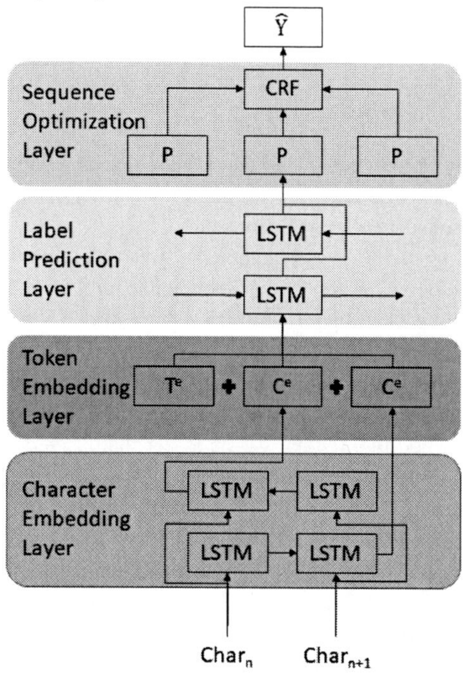

Figure 1: The character-enhanced bi-LSTM CRF system architecture. Where T^e is the token embedding, C^e are the character embeddings, and P is the bi-LSTM's predicted class.

4 Training

The hyperparameters that yield the best results were identified as: character embeddings with 25 dimensions, character level LSTM hidden states that use 25 dimensions, token embeddings with 100 dimensions, token level context embeddings with 1,024 dimensions, and a token level hidden state that uses 100 dimensions. We limited the model to 100 epochs with early stopping when the validation set's F1 score did not improve after 10 epochs. Early stopping was triggered when the model's F1 score on the validation set peaked then failed to achieve a better score within ten more epochs. We used a learning rate of 0.005. We clipped gradients at 5.0 and applied a dropout rate of 0.5. We tested several other hyperparameters with the model to include 200 dimension token embeddings, 2,048 context embeddings, 0.001 learning rate, 0.4 and 0.6 dropout rates. None of these provided significant increases in performance, however, some did cause large increases in training and inference times. Using a 16 core CPU, word embeddings are trained in 8 minutes and 43 seconds and training the model takes 19 minutes and 22 seconds. Due to the small data set size, only 3GB of free RAM is necessary to train the system.

5 Evaluation and Results

We measured performance of the system based on provided gold label AEs. We used Precision, Recall, and F1 Score to monitor a model's performance as it trained and to check that the reported values were reflective of the model's ability to generalize to the test set. Due to the inherently noisy nature of user generated social media text, we found that noise reduction techniques performed during the preprocessing stage had a much higher impact on model performance than hyperparameter tuning. Swapping tokenizers netted performance increases in F1 Score as big as 9.73, when keeping special characters, and 8.07, when not. Table 1 shows that best results on the test set are achieved with NLTK's Word Punct tokenizer and when special characters are kept.

Tokenizers	Special Characters	P	R	F1
Stanford	Yes	35.12%	47.19%	40.27
Stanford	No	33.13%	52.27%	40.55
NLTK Word	Yes	42.08%	58.17%	48.83
NLTK Word	No	**46.09%**	51.46%	48.62
Word Punct	Yes	42.79%	**60.13%**	**50.00**
Word Punct	No	39.52%	53.25%	45.27
Whitespace	Yes	32.90%	58.22%	42.04
Whitespace	No	44.29%	52.92%	48.22

Table 1: System performance on test set with different tokenizers.

The shared task was evaluated using a total of six performance metrics including both strict and relaxed variants of Precision, Recall, and F1 Score. Table 2 shows that our final system provided a 59.7 Relaxed F1 Score and a 40.7 Strict F1 Score on the evaluation set, beating shared task averages by 5.9 and 9.0, respectively.

Metric	Our System	Task Average
Relaxed Precision	59.6%	51.3%
Relaxed Recall	59.9%	61.7%
Relaxed F1 Score	59.7	53.8
Strict Precision	40.6%	30.3%
Strict Recall	40.7%	35.8%
Strict F1 Score	40.7	31.7

Table 2: System Performance on evaluation set.

Error analysis shows that words heavily associated with AEs, such as "withdrawal", are almost always accurately identified as being AEs. Alternatively, words with neither positive nor negative connotations are frequently missed as being AEs, such as "sleep" in "it could be two months before i sleep well again". Errors also occurred when tokens frequently associated with AEs were present but not in relation to medication usage. An example would be the identification of "rejection hurts" in "rejection hurts, cymbalta can help". The model appears to give excessive weight to the specific word being used while not giving enough weight to the word's context. Future work would explore the use of a larger corpus that includes more negative examples of those words, additional LSTM layers in the label prediction layer, and the use of more recent word embedding algorithms.

References

Franck Dernoncourt, Ji Young Lee, Ozlem Uzuner, and Peter Szolovits (2017). De-identification of patient notes with recurrent neural networks. *Journal of the Amerian Medical Informatics Association, volume 24* (Issue 3), 596-606.

Sepp Hochreiter and Jurgen Schmidhuber (1997). Long Short-term Memory. *Neural Computation, volume 9* (Issue 8), 1735–1780.

Edward Loper, Ewan Klein, and Steven Bird (2009). *Natural Language Processing with Python.* O'Reilly Media.

Christopher D. Manning, Mihai Surdeanu, John Bauer, Jenny Finkel, Steven J. Bethard, and David McClosky (2014). The Stanford CoreNLP Natural Language Processing Toolkit. In *Proceedings of 52nd Annual Meeting of the Association for Computational Linguistics: System Demonstrations*, 55-60.

Jeffrey Pennington, Richard Socher, and Christopher Manning (2014). Glove: Global Vectors for Word Representation. In *Proceedings of the 2014 Conference on Empirical Methods in Natural Language Processing*, 1532-1543.

Matthew E. Peters, Mark Neumann, Mohit Iyyer, Matt Gardner, Christopher Clark, Kenton Lee, and Luke Zettlemoyer (2018). Deep Contextualized Word Representations. In *Proceedings of the 2018 Conference of the North American Chapter of the Association for Computational Linguistics: Human Language Technologies, volume 1.*

Davy Weissenbacher, Abeed Sarker, Arjun Magge, Ashlynn Daughton, Karen O'Connor, Michael Paul, and Graciela Gonzalez-Hernandez (2019). Overview of the Fourth Social Media Mining for Health (SMM4H) Shared Task at ACL 2019. In *Proceedings of the 2019 ACL Workshop SMM4H: The 4th Social Media Mining for Health Applications Workshop & Shared Task.*

Using machine learning and deep learning methods to find mentions of adverse drug reactions in social media

Pilar López-Úbeda, Manuel Carlos Díaz-Galiano,
Maria-Teresa Martín-Valdivia, L. Alfonso Ureña-López

Department of Computer Science, Advanced Studies Center in ICT (CEATIC)
Universidad de Jaén, Campus Las Lagunillas, 23071, Jaén, Spain
{plubeda, mcdiaz, maite, laurena}@ujaen.es

Abstract

Today social networks play an important role, where people can share information related to health. This information can be used for public health monitoring tasks through the use of Natural Language Processing (NLP) techniques. Social Media Mining for Health Applications (SMM4H) provides tasks such as those described in this document to help manage information in the health domain.

This document shows the first participation of the SINAI group in SMM4H. We study approaches based on machine learning and deep learning to extract adverse drug reaction mentions from highly informal texts in Twitter.

The results obtained in the tasks are encouraging, we are close to the average of all participants and even above in some cases.

1 Introduction

An Adverse Drug Reaction (ADR) is an injury occurring after a drug (medication) is used at the recommended dosage, for recommended symptoms. This is a area that has already been researched in recent years (Sarker and Gonzalez, 2015; Karimi et al., 2015), and in which we will contribute with new systems.

The proposed shared tasks of SMM4H continue with NLP challenges in social media mining for health monitoring and surveillance (ws-, 2018; Weissenbacher et al., 2018).

We have decided to participate in 2 of the 4 tasks proposed by the organizers: automatic classifications of adverse effects mentions in tweets and extraction of adverse effect mentions.

In task *automatic classifications of adverse effects* the goal is a binary classification problem. The designed system for this sub-task should be able to distinguish tweets reporting an Adverse Effect (AE) from those that do not.

In the second task called *Extraction of Adverse Effect mentions*. This task includes identifying the text span of the reported ADRs and distinguishing ADRs from similar non-ADR expression. ADRs are multi-token, descriptive, expressions, so this subtask requires advanced Named Entity Recognition (NER) approaches.

2 Tweet data

The corpus are composed of tweets extracted from the famous social network called Twitter. This social network allows people to freely post short messages (called tweets) of up to 140 characters. Twitter has rapidly gained popularity worldwide, with more than 326 million active users generating more than 500 million tweets daily.

- Data set for task 1: For each tweet, the publicly available data set contains: (i) the user ID, (ii) the tweet ID, and (iii) the binary annotation indicating the presence or absence of ADRs.

  The training data is composed of 25,672 tweets (2,374 positive and 23,298 negative) and the test data contains 4,5175 tweets.

- Data set for task 2: This set contains a subset of the tweets from Task 1 tagged as *hasADR* plus an equal number of *noADR* tweets. The corpus contains: (i) the tweet ID, (ii) the start and (iii) end of the span, (iv) the annotation indicating an ADR or not and (v) the text covered by the span in the tweet.

  The training data is composed of 2,367 tweets (1,212 positive and 1,155 negative) and the test data contains 1,573 tweets.

3 Taking part in tasks

In this section we will explain the 3 methodologies applied to each task.

Proceedings of the 4th Social Media Mining for Health Applications (#SMM4H) Workshop & Shared Task, pages 102–106
Florence, Italy, August 2, 2019. ©2019 Association for Computational Linguistics

Before beginning to implement our approaches, it is necessary to clean the text of some rare characters that we find, these characters can make noise to our systems, therefore, we must treat them correctly. This pre-processing has been:

- Convert the text to lowercase.

- Substitution of characters HTML like: *&*, *<*, and *>* to your representation: *&*, *<* and *>*.

3.1 Task 1: Automatic classifications of adverse effects mentions in tweets

In addition to the text processing already carried out and described above, for this task we have also decided to carry out another pre-processing:

- Expand contractions: the contractions in the text have been expanded as for example: *you're* to *you are*

- Remove hashtag: for this task we consider that the hashtag add noise to the text as we do not process them.

- Remove @ mentions: mentions of persons have been removed from the text.

- Remove non-alphanumeric words: we have only taken into account alphanumeric words.

For Task 1 systems we have used the automatic learning and deep learning approaches described below:

3.1.1 SVM

SVM (Vector Support Machines) is one of the best classifiers for a wide range of situations, so it is considered one of the references within the field of statistical learning and machine learning. We used SVM with linear kernel.

For tweet processing we have applied the TF-IDF schema with the following parameters: min_df = 3, max_df = 0.8, sublinear_tf = True, use_idf = True, lowercase = True and ngram_range = (1,3).

This will be our baseline, from which we will depart for better results.

3.1.2 SVM + features

For this system, we have used the SVM of the previous baseline adding some relevant features for this specific task. We believe it is interesting to use external resources referring to the medical domain.

We have used the medical entity recognizer for English called MetaMap (Aronson, 2001). MetaMap is a widely available program providing access to the concepts in the unified medical language system (UMLS[1]) Metathesaurus from biomedical text. In addition, this resource provides additional information about the medical concept detected. For example, we can know the Concept Unique Identifier (CUI), the preferred name or the semantic type for the concept.

We make use of the semantic type of the concepts detected, and specifically, we use the semantic groups: "dsyn", "fndg", "inpo", "menp", "mobd", "neop", "patf", "phsf", "sosy", "topp" creating a vector of 10 positions, we insert 1 in the case in which it finds a concept in the tweet with that semantic group, 0 in other cases.

These semantic groups can be understood as: Disease or Syndrome, Finding, Injury or Poisoning, Mental Process, Mental or Behavioral Dysfunction, Neoplastic Process, Pathologic Function, Physiologic Function, Sign or Symptom and Therapeutic or Preventive Procedure respectively.

We decided to use these semantic types thanks to the ADR mentioned in Task 2 corpus, these ADR were introduced in MetaMap and we chose to use the 10 most repeated semantic groups.

3.1.3 CNN

For the third system, we implemented a Convolutional Neural Network (CNN). CNN are a category of neural networks that have proven very effective in areas such as image recognition and classification.

The architecture of the network is as follows:
- Embedding layer.
- 1D convolution layer: filters = 32, convolution window = 3, activation = relu and the other default values.
- 1D Max pooling layer: size of the max pooling windows = 2 and the other default values.
- 1D convolution layer: filters = 32, convolution window = 3, activation = relu and the other default values.
- Global max pooling layer with default values.
- Dense layer for output with 1 output unit and activation = sigmoid.

We have used the Twitter pre-trained word vec-

---

[1] `https://www.nlm.nih.gov/research/umls/`

tors of GloVe[2]. These embeddings are composed of 2B tweets, 27B tokens, 1.2M vocab and 200 dimension.

3.2 Task 2: Extraction of Adverse Effect mentions

In the second task, our team has focused on the use of Conditional Random Field (CRF) algorithm, applying characteristics to it in such a way that they provide extra information to each word of the document.

3.2.1 CRF

CRF classifier is a stochastic model commonly used to label and segment data sequences or extract information from documents. We used CRFsuite, the implementation provided by Okazaki, as it is fast and provides a simple interface for training/modifying the input features.

The CRF classifier is trained on annotated mentions of ADRs and indications, and it attempts to classify individual tokens in sentences. Therefore, it learns to distinguish five different labels: ADR and 0.

Below, we define some characteristics for each word in the document used in all our models:

- Characteristics of the context: Context is defined by three characteristics that include the current word (word), the previous word (word-1) and the subsequent word in the sentence (word+1).

- POS: Part of speech of the token, which was generated using the Spacy[3] library for Python.

- Lemma: Lemma of the token, which was generated using the Spacy.

- Other features: we incorporate some basic features of each word such as isLower, isUpper, isTitle, isDigit, isAlpha, isBeginOfSentence and isEndIfSentece.

3.2.2 CRF + W2V

We want to use embedded word vectors as feature in existing conditional random field (CRF) with gazetteer features for sequence labeling task in text.

We have again used the Twitter pre-trained word vectors of GloVe but with 50 dimension.

To make this possible, we added 50 new features to each word, to the previous word and to the next word. These 50 characteristics refer to each

---

[2] https://nlp.stanford.edu/projects/glove/

[3] https://spacy.io/

Bitchain	Word	Count
011110100000	unmotivated	754
011110100000	knackered	2407
011110100000	tired	232683
011110100000	exhausted	19368
011110100000	drained	3333

Table 1: Example content of Brown cluster.

of the dimensions of that word. In this way the algorithm will learn where the words are within the axes in order to improve in context.

3.2.3 CRF + BC + W2V

For the last system developed for this task, the word representations feature induced by Brown clustering method was introduced as an additional feature.

Brown clustering (Brown et al., 1992) is a greedy, hierarchical, agglomerative hard clustering algorithm to partition a vocabulary into a set of clusters with minimal loss in mutual information. Intuitively, the Brown clustering method will merge the tokens with similar contexts into the same cluster.

The implementation of Brown clustering method by Liang and described by Owoputi et al. is adopted in our system. The clustering used contains 216,846 words, is grouped in 1000 clusters and processed more than 56 million tweets.

Some examples of Brown clustering are shown in Table 1. In this table we can see how different words are in the same cluster (011110100000) and the number of occurrences found.

The feature that was finally added to the method was the bitchain to which each word belonged.

4 Results

In this section we show the results obtained by the group SINAI in the participation of SMM4H Shared Task 2019.

4.1 Task 1

The average of all participants in Task 1 and the results obtained by our group in Task 1 are those shown in Table 2.

As we can see the mean has a low measure, so we can intuit that it is a difficult task. In our case, the neural network learns better than machine learning systems, although we add features

Approach	F1	Prec	Recall
Average particp.	**0.5019**	0.5351	**0.5054**
SVM	0.4509	**0.6393**	0.3482
SVM + features	0.4829	0.6222	0.3946
CNN	0.4969	0.5517	0.4521

Table 2: Result obtained for Task 1.

Approach	F1	Prec	Recall
Average particp.	0.5383	0.5129	**0.6174**
CRF	0.496	0.633	0.408
CRF + W2V	0.532	**0.616**	0.468
CRF + BC + W2V	**0.542**	0.612	0.486

Table 3: Result obtained for Task 2 relaxed matching.

to these models.

Although the use of features added to SVM improves our baseline in F1 and recall, they are not sufficient and we do not get a substantial increase. We can observe that systems 2 and 3 worsen the precision. For future work we can try to choose some features more related to the task.

4.2 Task 2

In this task two measures of agreement were computed: strict and relaxed matching.

- Relaxed matching

  The average scores for this task with relaxed matching and our results are showing in Table 3.

  In different measures such as F1 and precision we are above average. In terms of precision, we exceeded it by 20%, although the average recall does not reach it and that hurts us.

- Strict matching

  Our results and the average scores for all participants in this task with strict matching are presented in Table 4.

  In this system, we can see that the same thing happens as in the case of relaxed matching, we surpass the F1 and precision measures, but not in recall. For next participation we will pay special interest in the exhaustiveness for relevant instances that we have recovered.

We will be able to analyze the results once the organizers provide us with the complete test. With

Approach	F1	Prec	Recall
Average particp.	0.3169	0.3026	**0.3581**
CRF	0.326	0.419	0.267
CRF + W2V	0.352	0.408	0.31
CRF + BC + W2V	**0.36**	**0.408**	0.322

Table 4: Result obtained for Task 2 strict matching.

this, we will be able to carry out an analysis of errors and see the failures obtained and how to improve them.

5 Conclusions

In this document, we expose the first participation of the SINAI group in SMM4H, we created 3 strategies for Task 1 and 3 strategies for Task 2. For Task 1 different approaches of machine learning and deep learning were implemented, whereas for Task 2 the effectiveness of several classification characteristics was explored in the training of the CRF model and it was found that context and cluster integration were the most contributing characteristics.

In both tasks we managed to overcome our baseline and improve in each method. In Task 1 we get a F1 of 0.486 being a little below the average of all participants, in Task 2 we managed to obtain a measure F1 of 0.322 in the strict system and 0.486 in relaxed system.

Our future work will involve exploring the effectiveness of training a deep learning neural network, rather than the CRF, to learn features and classify labels and improve our neural networks and add new text features. As well as participate in all tasks proposed to implement our systems and expose them to the scientific community.

Acknowledgments

This work has been partially supported by Fondo Europeo de Desarrollo Regional (FEDER) and REDES project (TIN2015-65136-C2-1-R) from the Spanish Government.

References

2018. *Proceedings of the 2018 EMNLP Workshop SMM4H: The 3rd Social Media Mining for Health Applications Workshop & Shared Task*. Association for Computational Linguistics, Brussels, Belgium.

Alan R Aronson. 2001. Effective mapping of biomedical text to the umls metathesaurus: the metamap program. In *Proceedings of the AMIA Symposium*, page 17. American Medical Informatics Association.

Peter F Brown, Peter V Desouza, Robert L Mercer, Vincent J Della Pietra, and Jenifer C Lai. 1992. Class-based n-gram models of natural language. *Computational linguistics*, 18(4):467–479.

Sarvnaz Karimi, Chen Wang, Alejandro Metke-Jimenez, Raj Gaire, and Cecile Paris. 2015. Text and data mining techniques in adverse drug reaction detection. *ACM Computing Surveys (CSUR)*, 47(4):56.

Percy Liang. 2005. *Semi-supervised learning for natural language*. Ph.D. thesis, Massachusetts Institute of Technology.

Naoaki Okazaki. 2007. Crfsuite: a fast implementation of conditional random fields (crfs).

Olutobi Owoputi, Chris Dyer, Kevin Gimpel, and Nathan Schneider. 2012. Part-of-speech tagging for twitter: Word clusters and other advances.

Abeed Sarker and Graciela Gonzalez. 2015. Portable automatic text classification for adverse drug reaction detection via multi-corpus training. *Journal of biomedical informatics*, 53:196–207.

Davy Weissenbacher, Abeed Sarker, Michael J. Paul, and Graciela Gonzalez-Hernandez. 2018. Overview of the third social media mining for health (SMM4H) shared tasks at EMNLP 2018. In *Proceedings of the 2018 EMNLP Workshop SMM4H: The 3rd Social Media Mining for Health Applications Workshop and Shared Task*, pages 13–16, Brussels, Belgium. Association for Computational Linguistics.

Towards text processing pipelines to identify adverse drug events-related tweets: University of Michigan @ SMM4H 2019 Task 1

V.G.Vinod Vydiswaran,[1,2*] Grace Ganzel,[2] Bryan Romas,[2] Deahan Yu,[2]
Amy M. Austin,[2] Neha Bhomia,[2] Socheatha Chan,[2] Stephanie V. Hall,[1] Van Le,[2]
Aaron Miller,[2] Olawunmi Oduyebo,[2] Aulia Song,[2] Radhika Sondhi,[2] Danny Teng,[2]
Hao Tseng,[2] Kim Vuong,[2] Stephanie Zimmerman[2]

[1]Department of Learning Health Sciences, [2]School of Information, University of Michigan
*Corresponding author: vgvinodv@umich.edu

Abstract

We participated in Task 1 of the Social Media Mining for Health Applications (SMM4H) 2019 Shared Tasks on detecting mentions of adverse drug events (ADEs) in tweets. Our approach relied on a text processing pipeline for tweets, and training traditional machine learning and deep learning models. Our submitted runs performed above average for the task.

1   Introduction

A growing number of users produce and share information on the internet, including health information. As of 2017, the number of social media users increased by approximately one million per day, with approximately half of adults worldwide using some form of social media (Kemp, 2017). According to (Domo, 2018), 473,400 tweets were sent every minute in 2018.

The discussion of health-related information on social media is becoming increasingly common, which can be utilized by researchers in a multitude of ways, including pharmacovigilance and public health surveillance (Nikfarjam et al., 2015). Numerous works have utilized tweets to analyze public health concerns, with many focusing specifically on the identification of adverse drug reactions (ADRs) (O'Connor et al., 2014). Tweets can contain important information, including the mention of specific medications, indications for use, and side effects. Additional information can also be obtained, such as time of the tweet, location, and user characteristics (Paul and Dredze, 2011).

The Social Media Mining for Health Applications (SMM4H) Shared Tasks were organized to provide labeled social media data sets for natural language processing (NLP) researchers to study challenges in health monitoring and surveillance (Weissenbacher et al., 2019). Task 1 focused on classification of adverse drug event (ADE) mentions in tweets, where the participating systems were expected to distinguish tweets reporting an ADE from those that do not, taking into account subtle linguistic variations between adverse effects and indications, such as the reason to use the medication.

2   Data cleaning and pre-processing

Our approach was to develop a text processing pipeline to clean and process tweets and identify tweets that mention ADEs. We ran the following pre-processing steps:

(a) Removing UTF-8 characters: All UTF-8 characters in the tweets were removed or replaced with relevant tags. For example, a pill emoji was replaced with the tag '⟨pill⟩', and a dizzy-faced emoji was replaced with the tag '⟨dizzy⟩'.

(b) Running Ekphrasis: After all UTF-8 characters were removed, the Ekphrasis text processing tool (Baziotis et al., 2017) was run with the following minor modifications. First, because the tool was unable to unpack contractions that appeared in uppercase text, regular expressions were written to capture all uppercase tokens for manual verification and tagging. After tagging, the tweets were converted to lowercase, allowing them to be fully processed by the unpacking feature. New contractions were added to the Ekphrasis unpacking routine based on a manual review of Ekphrasis output, when applied to the challenge data set.

(c) NLTK TweetTokenizer and Lemmatizer: NLTK TweetTokenizer was run to further process the tweets, and the outputted tweets were then lemmatized.

(d) MetaMap: Each tweet was run through MetaMap, and concept and semantic types identified within the text with a MetaMap score above 800 were extracted as features.

Proceedings of the 4th Social Media Mining for Health Applications (#SMM4H) Workshop & Shared Task, pages 107–109
Florence, Italy, August 2, 2019. ©2019 Association for Computational Linguistics

(e) cTAKES: Tweets were run through cTAKES to identify concepts from the Systematized Nomenclature of Medicine (SNOMED). The identified SNOMED codes were added as features.

(f) Pattern-based features: Additional features were generated based on pattern-matching rules using regular expressions.

3 Word representation for neural models

We trained two variations of neural network models — a bidirectional LSTM (Graves and Schmidhuber, 2005) model, and a bidirectional LSTM model with a convolutional neural network (CNN) layer (Kim, 2014). We also compared the performance of both models using pre-trained GloVe word embedding (Pennington et al., 2014) and using pre-trained Word2vec Twitter word embedding (Godin et al., 2015). To evaluate the performance of the models, we randomly split the training set in a 80-20 ratio while maintaining the original class proportions. The four models were trained on 80% of the provided training data and tested on the remaining 20% (validation data set). Of the four models, the best model based on validation accuracy was chosen as the final model, which was the bidirectional LSTM model using GloVe word embedding.

3.1 Features

To generate the input tweet representation for deep learning models, we undertook the following additional steps:

(a) Part-of-speech tag embedding: To create a part-of-speech (POS) embedding, we used NLTK to first extract POS labels for each word. We then converted each tweet into a sequence of POS tags according to the token order and created the POS tag embedding.

(b) First-character embedding: Similar to the part-of-speech tag embedding, we extracted the first character of each token in a tweet and generated four binary features depending on whether the first character was an uppercase letter, a lowercase letter, an integer, or a symbol / special character.

(c) Medical dictionary: Finally, we obtained a MedDRA dictionary from Side Effect Resource, SIDER (Kuhn et al., 2016, 2010) and used it to create a one-hot vector representation for words listed in SIDER, in addition to word embedding.

4 Description of runs

Once the pre-processing and input representation were finalized, we trained the following three models corresponding to the three submitted runs:

Run 1: As a baseline for our models, we trained a linear kernel support vector machine classifier with balanced class weights. The model was trained over unigram features generated from the lowercased tweet text. The individual feature weights were computed using their inverse document frequency over the training data set. The classifier was built using scikit-learn.

Run 2: For the second run, we ran all tweets through the pre-processing pipeline described in Sec. 2. The tweet text was cleaned using the modified Ekphrasis tool, features from MetaMap and cTAKES were added, and the text was tokenized. Unigram and bigram features were instantiated and were weighted by the inverse document frequency in the training set. A linear kernel support vector machine classifier was trained with a balanced class weight configuration.

Run 3: For the third run, we used a bidirectional LSTM with categorical cross entropy loss function with RMSprop optimizer. We set the model dropout layer probability to 0.2 in order to avoid overfitting. Following (Vaswani et al., 2017), we added an attention layer. Our output layer for the classification task was a dense layer followed by the softmax function. For the input representation, we employed a concatenation of the pre-trained GloVe word embedding and the first character embedding. We padded each tweet to 29 tokens, which is the sum of the average tweet length ($\ell = 16$) and two standard deviations of the tweet lengths ($\sigma = 6.5$) in the 80% data set. We set it this way because the maximum length from the 80% of the data was too long ($\ell = 130$ tokens) to use and the average length was too short to cover substantial amount of tweets. The model was trained on the 100% of the provided data (both training and validation sets) and run for 100 epochs.

5 Results

In all, 16 teams participated in Task 1 for a total of 43 runs. Table 1 summarizes the performance of our three runs and the average over all runs submitted to the task. Runs 1 and 2 were better than the average performance over recall and F1 mea-

Run ID	Pred pos (%)	Prec	Rec	F1
Run 1	865 (18.9%)	0.452	**0.625**	0.525
Run 2	566 (12.4%)	**0.565**	0.511	**0.537**
Run 3	492 (10.8%)	0.555	0.436	0.488
Avg. task performance		0.535	0.505	0.502

Table 1: Performance of the submitted runs in terms of count (and percentage) of predicted positive instances, precision, recall, and F1 over the test set (n = 4,575).

sures, while runs 2 and 3 were better than the average run on precision. Run 2 was the best among the three submitted runs. It identified 566 (12.4%) tweets as positive with a precision of 0.565, recall of 0.511, and F1 measure of 0.537. All these measures were better than the average measures among runs submitted for Task 1.

6 Conclusion

Our approach for participating in the 2019 SMM4H Shared Task 1 was to develop a text processing pipeline for tweets, focusing on pre-processing, feature weighting, and training traditional feature-based and deep learning models. Our runs performed above the average shared task performance, and the best run achieved an F1 measure of 0.537. Additional runs are planned to further analyze the performance of deep learning models on this task.

References

Christos Baziotis, Nikos Pelekis, and Christos Doulkeridis. 2017. DataStories at SemEval-2017 Task 4: Deep LSTM with attention for message-level and topic-based sentiment analysis. In *Proceedings of the 11th International Workshop on Semantic Evaluation (SemEval-2017)*, pages 747–754, Vancouver, Canada. ACL.

Domo. 2018. Data never sleeps 6.0. Retrieved from https://www.domo.com/learn/data-never-sleeps-6.

Fréderic Godin, Baptist Vandersmissen, Wesley De Neve, and Rik Van de Walle. 2015. Multimedia Lab @ ACL WNUT NER Shared Task: Named entity recognition for Twitter microposts using distributed word representations. In *Proceedings of the Workshop on Noisy User-generated Text*, pages 146–153, Beijing, China. ACL.

Alex Graves and Jurgen Schmidhuber. 2005. Framewise phoneme classification with bidirectional lstm networks. In *Proceedings of the IEEE International Joint Conference on Neural Networks*, volume 4, pages 2047–2052.

Simon Kemp. 2017. The global state of the Internet in April 2017. Retrieved from https://thenextweb.com/contributors/2017/04/11/current-global-state-internet/.

Yoon Kim. 2014. Convolutional neural networks for sentence classification. In *Proceedings of the 2014 Conference on Empirical Methods in Natural Language Processing, EMNLP 2014, October 25-29, 2014, Doha, Qatar, A meeting of SIGDAT, a Special Interest Group of the Association for Computational Linguistics*, pages 1746–1751.

Michael Kuhn, Monica Campillos, Ivica Letunic, Lars Juhl Jensen, and Peer Bork. 2010. A side effect resource to capture phenotypic effects of drugs. *Molecular Systems Biology*, 6:343.

Michael Kuhn, Ivica Letunic, Lars Juhl Jensen, and Peer Bork. 2016. The SIDER database of drugs and side effects. *Nucleic Acids Research*, 44(Database issue):D1075–D1079.

Azadeh Nikfarjam, Abeed Sarker, Karen O'Connor, Rachel Ginn, and Graciela Gonzalez. 2015. Pharmacovigilance from social media: Mining adverse drug reaction mentions using sequence labeling with word embedding cluster features. *Journal of the American Medical Informatics Association : JAMIA*, 22(3):671–681.

Karen O'Connor, Pranoti Pimpalkhute, Azadeh Nikfarjam, Rachel Ginn, Karen L Smith, and Graciela Gonzalez. 2014. Pharmacovigilance on twitter? mining tweets for adverse drug reactions. In *Proceedings of the AMIA Annual Symposium*, pages 924–933.

Michael J. Paul and Mark Dredze. 2011. You are what you tweet: Analyzing twitter for public health. In *Proceedings of the Fifth International AAAI Conference on Weblogs and Social Media*, pages 265–272.

Jeffrey Pennington, Richard Socher, and Christopher Manning. 2014. Glove: Global vectors for word representation. In *Proceedings of the 2014 conference on empirical methods in natural language processing (EMNLP)*, pages 1532–1543.

Ashish Vaswani, Noam Shazeer, Niki Parmar, Jakob Uszkoreit, Llion Jones, Aidan N Gomez, Łukasz Kaiser, and Illia Polosukhin. 2017. Attention is all you need. In I. Guyon, U. V. Luxburg, S. Bengio, H. Wallach, R. Fergus, S. Vishwanathan, and R. Garnett, editors, *Advances in Neural Information Processing Systems 30*, pages 5998–6008. Curran Associates, Inc.

Davy Weissenbacher, Abeed Sarker, Arjun Magge, Ashlynn Daughton, Karen O'Connor, Michael Paul, and Graciela Gonzalez-Hernandez. 2019. Overview of the fourth Social Media Mining for Health (SMM4H) shared task at ACL 2019. In *Proceedings of the 2019 ACL Workshop SMM4H: The 4th Social Media Mining for Health Applications Workshop & Shared Task*.

Neural Network to identify personal health experience mention in tweets using BioBERT embeddings

Shubham Gondane[*]
Arizona State University
`sgondane@asu.edu`

Abstract

This paper describes the system developed by team ASU-NLP for the Social Media Mining for Health Applications(SMM4H) shared task 4. We extract feature embeddings from the BioBERT (Lee et al., 2019) model which has been fine-tuned on the training dataset and use that as inputs to a dense fully connected neural network. We achieve above average scores among the participant systems with the overall F1-score, accuracy, precision, recall as 0.8036, 0.8456, 0.9783, 0.6818 respectively.

1 Introduction

There has been an increase in the use of social media worldwide in recent years, which provides an abundance of data available and an exciting opportunity to build and improve biomedical and public health applications. The Social Media Mining for Health Applications (SMM4H) Workshop 2019 (Weissenbacher et al., 2019) proposed four tasks. We have focused on task 4, which was the most interesting. The task is to classify whether the tweet contains personal health mention as opposed to a general discussion of the topic. The training data consisted of tweets related to the flu. The system is evaluated on tweets related to flu and a second health domain across two contexts.

1.1 Data Description

The organizers provided two datasets across different contexts, but both in the flu domain. The first dataset had 1046 records of flu infection, but around 1023 tweets were available for download. The flu vaccination dataset had around 9800 records out of which only 6659 were available for download. The combined dataset had 7682 tweets in total.

---

[*] The author is advised by Dr. Chitta Baral at Arizona State University.

1.2 Related Work

Much previous work has focused on tracking and monitoring diseases on social media. Identifying various health ailments in social media by (Paul and Dredze, 2011) introduced a topic model based system using LDA to discover health mentions. Previous work done on creating generalizable classifiers have used traditional machine learning based approaches. (Yin et al., 2015) have developed a scalable system by training classifiers on a dataset of 34 health topics. They created a general health classifier using standard SVM with an accuracy of 77 percent. More recently, (Karisani and Agichtein, 2018) developed a system called as WESPAD that combines lexical, syntactic, word embedding-based, and context-based features. The authors report that the system can generalize from a few examples by automatically distorting the word embedding space to detect the accurate health mentions most effectively.

1.3 Preprocessing

The challenge in this task is to train a model on one disease domain and test on another, so it is important to make sure the model does not learn disease-specific characteristics. One way to ensure this is to mask specific terms like flu or influenza mentions with an AILMENT tag. A list of all flu-related terms was created using a pretrained Word2Vec model for Twitter (Godin et al., 2015) to find similar terms to flu. The list was expanded using human knowledge and ConceptNet[1] (Speer et al., 2017). This list of terms was used to mask all the flu mentions in the dataset.

Additionally we use the preprocessing library Ekphrasis to clean the tweets. (Baziotis et al., 2017).

---

[1] `www.conceptnet.io`

Proceedings of the 4th Social Media Mining for Health Applications (#SMM4H) Workshop & Shared Task, pages 110–113
Florence, Italy, August 2, 2019. ©2019 Association for Computational Linguistics

- All @user mentions were replaced by @user tag.

- All HTTP URLs were replaced by URL tag.

- Hashtags were preprocessed by removing the # symbol and keeping the words.

- Emojis, dates, numbers, etc. are removed.

2 Experiments

Language models like BERT (Devlin et al., 2018) and OpenAI GPT-2 (Radford et al., 2019) have achieved state of the art performances in various NLP tasks. Such models that are trained on large datasets can be fine-tuned on smaller datasets to achieve good scores on various NLP tasks. BioBERT (Lee et al., 2019), a domain-specific language representation model designed for biomedical text, is built using BERT architecture. Our system is built using transfer learning approach by fine tuning on the given dataset using the BioBERT model.

2.1 Fine-tuning

The fine-tuning process involves creating a train and dev set in the format provided by the data processor in the BERT/ BioBERT model. The BERT-base uncased model is used for the experiments [2]. The model is then trained on a sentence classification task end to end using the default parameter values provided by the authors. Fine-tuning on smaller dataset results in a high variance in the dev set accuracy. So the model with the best result on the dev set is selected after five iterations of the fine-tuning process. This process is applied for fine-tuning both the BERT and BioBERT v1.0 models. BioBERT produced a slightly better model with the difference in dev set accuracies of the final BERT and BioBERT fine-tuned models was less than 2 percent.

We also experimented with fine-tuning without doing any preprocessing on the tweets. As expected, the performance decreased quite significantly because BERT does token level masking and presence of URLs, hashtags, and @user-mentions makes this token level prediction more difficult.

2.2 Dense Neural Network Model

The BERT model can also be used for extracting features by fine-tuning the model and extracting the fixed contextual representations of each token. These features can be used in conjunction with other features in a different model. Fine-tuning is essential because the training set for these models is quite different from the dataset for this task. It helps to adjust the model weights that are closer to the target domain.

The BERT/BioBERT model adds two tokens in each input line - a CLS token in the beginning and SEP token at the end. Two feature embeddings are extracted in the following manner. In one case, we mask the flu-mentions, and in the other, the flu-mentions are kept as it is. The embedding for the CLS token is extracted by concatenating the weights of the last four layers of the BioBERT model. In their paper (Devlin et al., 2018), the authors state that concatenating last four layers gives the best result.

These embeddings are used as the input layer to a dense neural network with two hidden layers. We tried using these embeddings separately and also concatenated the two. The concatenated embedding performed slightly better than just using either of them separately. The final network has a 6144-dimensional input layer followed by two hidden layers of 512 and 128 dimensions, respectively. A dropout layer is added between the two hidden layers, and the hyperparameters are tuned accordingly.

3 Results and Discussion

Since the test set contained tweets related to undisclosed context we created a list of health concerns discussed on Twitter from previous research work (Daughton et al., 2018) (Paul and Dredze, 2014) (Dalrymple et al., 2016) (Khatua et al., 2019) done on exploring health-related tweets for analysis. This extensive list was used to mask the tweets of test set so that the masked embeddings make some contribution to the classification.

The system we used for this task shows that language models like BERT and BioBERT can be fine-tuned on a small dataset of tweets and still achieve promising results on test set where the health concern was similar to the training set. Transfer learning across different domains is still a challenging task as it is evident from the results.

It is interesting given that these models are

---

[2]The BioBERT model v1.0 used in this system is also based on the BERT-base model.

Model	Acc	F1	P	R
BERT fine-tuned without preprocessing	0.82	0.8	0.79	0.82
BERT fine-tuned	0.86	0.85	0.86	085
BioBERT fine-tuned	0.87	0.85	0.87	0.85
BioBERT unmasked embeddings	0.90	0.89	0.94	0.86
BioBERT masked embeddings	0.91	0.91	0.97	0.85
BioBERT masked and unmasked embeddings	**0.93**	**0.92**	**0.97**	**0.88**

Table 1: Accuracy, F1 score, Precision and Recall results on training data using different models and embeddings.

Test set	Acc	F1	P	R
health concern overall	0.84	0.80	0.97	0.68
health concern condition 1	0.92	0.92	0.98	0.86
health concern condition 2	0.69	0.51	0.91	0.35
health concern condition 3	0.80	0.59	1	0.42

Table 2: Final Accuracy, F1 score, Precision and Recall scores on the test set for the best performing run submitted.

trained on Wikipedia or biomedical text that how well they perform on tweets as tweets often contain misspellings, sarcasm, and slangs. It would also be interesting to see if the model can perform better if we had a BERT model trained on tweets or if we had a larger training dataset. This model could possibly be further improved by using additional data and the use of other textual and semantic features combined with the embeddings from the BioBERT model or trying different architectures.

References

Christos Baziotis, Nikos Pelekis, and Christos Doulkeridis. 2017. Datastories at semeval-2017 task 4: Deep lstm with attention for message-level and topic-based sentiment analysis. In *Proceedings of the 11th International Workshop on Semantic Evaluation (SemEval-2017)*, pages 747–754, Vancouver, Canada. Association for Computational Linguistics.

Kajsa E Dalrymple, Rachel Young, and Melissa Tully. 2016. facts, not fear negotiating uncertainty on social media during the 2014 ebola crisis. *Science Communication*, 38(4):442–467.

Ashlynn R Daughton, Michael J Paul, and Rumi Chunara. 2018. What do people tweet when theyre sick? a preliminary comparison of symptom reports and twitter timelines. In *ICWSM Social Media and Health Workshop*.

Jacob Devlin, Ming-Wei Chang, Kenton Lee, and Kristina Toutanova. 2018. Bert: Pre-training of deep bidirectional transformers for language understanding. *arXiv preprint arXiv:1810.04805*.

Fréderic Godin, Baptist Vandersmissen, Wesley De Neve, and Rik Van de Walle. 2015. Multimedia lab @ ACL WNUT NER shared task: Named entity recognition for twitter microposts using distributed word representations. In *Proceedings of the Workshop on Noisy User-generated Text*, pages 146–153, Beijing, China. Association for Computational Linguistics.

Payam Karisani and Eugene Agichtein. 2018. Did you really just have a heart attack?: towards robust detection of personal health mentions in social media. In *Proceedings of the 2018 World Wide Web Conference on World Wide Web*, pages 137–146. International World Wide Web Conferences Steering Committee.

Aparup Khatua, Apalak Khatua, and Erik Cambria. 2019. A tale of two epidemics: Contextual word2vec for classifying twitter streams during outbreaks. *Information Processing & Management*, 56(1):247–257.

Jinhyuk Lee, Wonjin Yoon, Sungdong Kim, Donghyeon Kim, Sunkyu Kim, Chan Ho So, and Jaewoo Kang. 2019. Biobert: a pre-trained biomedical language representation model for biomedical text mining. *arXiv preprint arXiv:1901.08746*.

Michael J. Paul and Mark Dredze. 2011. You are what you tweet: Analyzing twitter for public health. In *ICWSM*.

Michael J Paul and Mark Dredze. 2014. Discovering health topics in social media using topic models. *PloS one*, 9(8):e103408.

Alec Radford, Jeffrey Wu, Rewon Child, David Luan, Dario Amodei, and Ilya Sutskever. 2019. Language models are unsupervised multitask learners. *OpenAI Blog*, 1:8.

Robert Speer, Joshua Chin, and Catherine Havasi. 2017. Conceptnet 5.5: An open multilingual graph of general knowledge. In *Thirty-First AAAI Conference on Artificial Intelligence*.

Davy Weissenbacher, Abeed Sarker, Arjun Magge, Ashlynn Daughton, Karen O'Connor, Michael Paul, and Graciela Gonzalez-Hernandez. 2019. Overview of the fourth Social Media Mining for Health (SMM4H) shared task at ACL 2019. In *Proceedings of the 2019 ACL Workshop SMM4H: The 4th Social Media Mining for Health Applications Workshop & Shared Task*.

Zhijun Yin, Daniel Fabbri, S Trent Rosenbloom, and Bradley Malin. 2015. A scalable framework to detect personal health mentions on twitter. *Journal of medical Internet research*, 17(6):e138.

Give it a shot: Few-shot learning to normalize ADR mentions in Social Media posts

Manolis Manousogiannis
myTomorrows
Delft University of Technology
m.manousogiannis@mytomorrows.com

Sepideh Mesbah
Delft University of Technology
s.mesbah@tudelft.nl

Selene Baez Santamaria
myTomorrows

Alessandro Bozzon
Delft University of Technology

Robert-Jan Sips
myTomorrows

s.baez@mytomorrows.com a.bozzon@tudelft.nl r.sips@mytomorrows.com

Abstract

This paper describes the system that team MYTOMORROWS-TU DELFT developed for the 2019 Social Media Mining for Health Applications (SMM4H) Shared Task 3, for the end-to-end normalization of ADR tweet mentions to their corresponding MEDDRA codes. For the first two steps, we reuse a state-of-the-art approach, focusing our contribution on the final entity-linking step. For that we propose a simple Few-Shot learning approach, based on pre-trained word embeddings and data from the UMLS, combined with the provided training data. Our system (relaxed F1: 0.337-0.345) outperforms the average (relaxed F1 0.2972) of the participants in this task, demonstrating the potential feasibility of few-shot learning in the context of medical text normalization.

1 Introduction

Team MYTOMORROWS-TU DELFT participated in subtask 3 of the 2019 Social Media Mining for Health Applications (SMM4H) (Davy Weissenbacher, 2019) workshop, which is an end-to-end task. The goal is, given a tweet, to 1) automatically classify tweets containing an adverse drug reaction mention; 2) extract the exact ADR mention; 3) normalize the extracted ADR to its corresponding Medical Dictionary for Regulatory Activities (MEDDRA) code. The task is evaluated based on strict and relaxed F-score, precision and recall.

From an NLP perspective, this task poses a significant challenge as there is a large gap between the informal language used in social media and the formal medical language. Moreover, there is an absence of large annotated datasets, and datasets which are available often suffer from class imbalance. Illustrating this, Figure 1 provides an overview of the number of samples per class in the SMM4H task 3 dataset.

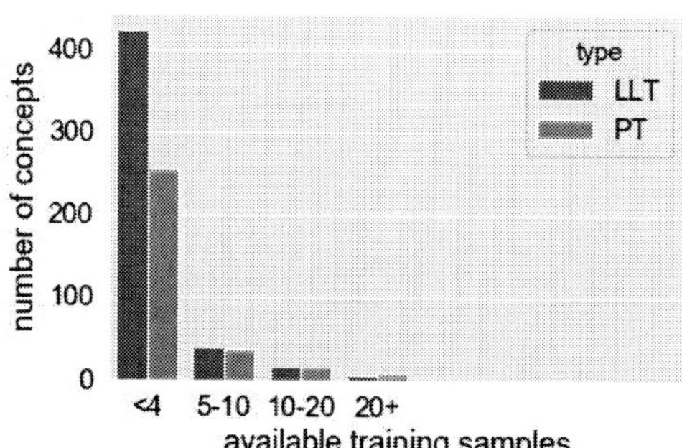

Figure 1: Available training samples per the medical concept present in the training data

Our end-to-end system consists of existing state-of-the-art for the first two steps. We focus our efforts on the third -normalization- step, which we formulate as a Few-Shot Learning problem (FSL), following the definition by Wang and Yao (Wang and Yao, 2019). In the following sections, we describe (1) the datasets that we worked on, (2) our approach in more detail and finally (3) our results and conclusions.

2 Data

2.1 Datasets

With the three subtasks, three manually annotated datasets were provided. All datasets contain tweets containing an ADR (positive) and without an ADR (negative). A brief overview of these datasets is provided in Table 1, but for more context we refer to (Davy Weissenbacher, 2019).

2.2 Preprocessing

The provided dataset for subtask 3 consists of ADR mentions, annotated with their corresponding MEDDRA code. In the hierarchy[1] of MEDDRA,

[1] https://www.meddra.org/how-to-use/basics/hierarchy

114

Proceedings of the 4th Social Media Mining for Health Applications (#SMM4H) Workshop & Shared Task, pages 114–116
Florence, Italy, August 2, 2019. ©2019 Association for Computational Linguistics

Task	Training data	
	#Positives	*#Negatives*
1	2374	23298
2	1212	1155
3	1212	1155

Table 1: Statistics of the training data used for task 1, 2 and 3

one Preferred Term (PT) is linked to one or more Lower Level Terms (LLTs) which are more specific descriptions of the related concept.

The provided dataset contains a mix of PTs and LLTs, mapping the 1212 ADR mentions to more than 500 different codes. Observing that the evaluation of the workshop task is performed on PT level, we map all annotations to the corresponding PT, as a preprocessing step. After this preprocessing step, the 1212 training mentions are mapped to 319 MEDDRA codes. Figure 1 provides an overview of the class distribution before and after preprocessing.

2.3 Prior Knowledge

In the training set for subtask 3, 149 out of the 319 MEDDRA codes that are present in the dataset (46.7%) have just one available training sample, while 254 (79.6%) have less than five training samples. To deal with the scarcity of samples, we create a prior knowledge dataset considering the 319 MEDDRA PTs in the training data. This dataset consists of the preferred names provided by the MEDDRA vocabulary and their corresponding preferred names in the Consumer Health Vocabulary (CHV), as mapped by the UMLS. The resulting dataset cointains 1,854 preferred names for the 319 MEDDRA codes.

3 Method

Our contributions focus on the normalization step, linking ADRs to their corresponding MEDDRA code. However, to be able to perform an end-to-end evaluation, we use existing state-of-the art techniques for subtask 1 (Sarker and Gonzalez, 2015) and 2 (Cocos et al., 2017), which we train on the workshop datasets [2].

The state-of-the-art approach for medical concept normalization in user-generated text is deep-

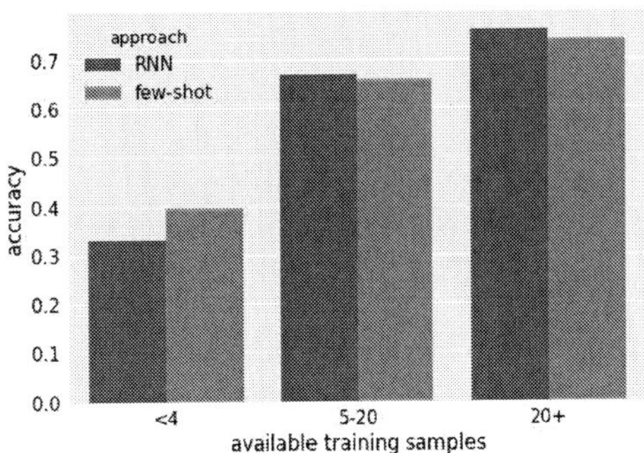

Figure 2: Accuracy per number of training samples.

neural networks (Limsopatham and Collier, 2016) which outperform traditional methods, when sufficient training data are available.

We trained both the CNN and RNN described by (Limsopatham and Collier, 2016) on the dataset for task 3, finding that the RNN has the best performance. On closer observation (and not surprisingly), we found that the accuracy of the RNN drops when fewer samples are available in the training data, as depicted in figure 2.

To deal with this drop in performance, we propose an embedding-based classifier that compares the ADR extracted mention to its 1-Nearest Neighbour on a vector space containing a) representations of the ADR mentions in the training data and b) representations of the prior knowledge dataset. Our intuition is that the embedding-based binary classifier would perform better on classes with a low number of samples, whereas an RNN would perform well on classes with higher sample numbers.

To create our embedding-based classifier we employ the pretrained Google News Word2Vec model (Mikolov et al., 2013). Using this model, we create vector representations for the ADR mentions in our training data[3]. Similarly we create vector representations for the mentions gathered in our prior knowledge dataset. At test time, we employ the same Word2Vec model to create a vector representation of the unseen ADR mention. Using a 1-Nearest Neighbour (with cosine similarity as distance metric), we then select the corresponding MEDDRA concept. Figure 2 shows that this model indeed seems less sensitive to low sample numbers.

---

[2]For task 1, we trained using the suggested settings, assigning 3:1 class weight favouring the ADR class. For task 2, we trained using the pre-trained-fixed setting.

[3]for mentions of more than one token we added the vectors

Technique	Relaxed			Strict		
	Precision	**Recall**	**F-score**	**Precision**	**Recall**	**F-score**
RNN	*0.318*	*0.337*	*0.327*	*0.232*	*0.246*	*0.239*
FSL	**0.336**	**0.355**	**0.345**	**0.237**	**0.252**	**0.244**
RNN+FSL (1)	*0.328*	*0.347*	*0.337*	*0.23*	*0.244*	*0.237*
RNN+FSL (2)	*0.331*	*0.35*	*0.34*	*0.235*	*0.249*	*0.242*
Task 3 AVG	*0.29*	*0.311*	*0.297*	*0.205*	*0.224*	*0.211*

Table 2: Relaxed and strict Precision/Recall/F-score for RNN, FSL, RNN+FSL (1) and (2) and the average score of all the participated team in task 3 (Task 3 AVG)

For our experiments, we use 4 systems: (1) RNN: the RNN proposed by (Limsopatham and Collier, 2016), trained on the both prior knowledge and the training set (which provides the best performance), (2) FSL: our 1-NN based on a combination of prior knowledge and the training set, (3) RNN+FSL (1): an ensemble of the RNN trained on only the training set and the FSL based on training + prior knowledge, and (4) RNN+FSL (2): an ensemble of the RNN trained on the training set and prior knowledge and the FSL based on training + prior knowledge. For our ensembles, we trust the model with the highest confidence (we used the cosine similarity for the 1-NN model to represent confidence) in case of disagreement.

4 Results

Our results are summarized in Table 2. Despite the fact that the RNN+FSL performed better in our development set, it did not generalize in the test data. On the test and evaluation data, FSL outperformed all the other techniques and achieved a 0.345 relaxed F-score and a 0.244 strict F-score which are above the average performance achieved in this task by all participants (i.e. Task 3 AVG).

5 Conclusions

In this paper, we describe our approach in subtask 3 of the SMM4II shared task for normalization of Adverse drug reaction mentions in Twitter posts. Our few-shot learning approach performs above the average in this task and hence we believe it to be a promising approach in cases where the amount of training data is limited.

As future work, we will focus on the discrimination between the ADRs that belong to one of the 'commonly seen cases' (classes with sufficient training data) from the 'rare cases' (classes with insufficient training data). This will allow us to efficiently combine a deep neural network with a few-shot learning approach into a more robust system that successfully links ADR tweet mentions into its MEDDRA codes.

References

Anne Cocos, Alexander G Fiks, and Aaron J Masino. 2017. Deep learning for pharmacovigilance: recurrent neural network architectures for labeling adverse drug reactions in twitter posts. *Journal of the American Medical Informatics Association*, 24(4):813–821.

Arjun Magge Ashlynn Daughton Karen O'Connor Michael Paul Graciela Gonzalez-Hernandez. Davy Weissenbacher, Abeed Sarker. 2019. Overview of the fourth social media mining for health (smm4h) shared task at acl 2019. in proceedings of the 2019 acl workshop smm4h: The 4th social media mining for health applications workshop shared task.

Nut Limsopatham and Nigel Collier. 2016. Normalising medical concepts in social media texts by learning semantic representation. In *Proceedings of the 54th Annual Meeting of the Association for Computational Linguistics (Volume 1: Long Papers)*, volume 1, pages 1014–1023.

Tomas Mikolov, Ilya Sutskever, Kai Chen, Greg S Corrado, and Jeff Dean. 2013. Distributed representations of words and phrases and their compositionality. In *Advances in neural information processing systems*, pages 3111–3119.

Abeed Sarker and Graciela Gonzalez. 2015. Portable automatic text classification for adverse drug reaction detection via multi-corpus training. *Journal of biomedical informatics*, 53:196–207.

Yaqing Wang and Quanming Yao. 2019. Few-shot learning: A survey. *arXiv preprint arXiv:1904.05046*.

BIGODM System in the Social Media Mining for Health Applications Shared Task 2019

Chen-Kai Wang[1], Hong-Jie Dai, PhD[2], Bo-Hung Wang[2]
[1]Big Data Laboratories, Chunghwa Telecom Laboratories, Taoyuan, Taiwan, R.O.C.
dennisckwang@gmail.com
[2]Department of Electrical Engineering,
National Kaohsiung University of Science and Technology, Kaohsiung, Taiwan, R.O.C.
{hjdai, 1061247131}@nkust.edu.tw

Abstract

In this study, we describe our methods to automatically classify Twitter posts conveying events of adverse drug reaction (ADR). Based on our previous experience in tackling the ADR classification task, we empirically applied the vote-based under-sampling ensemble approach along with linear support vector machine (SVM) to develop our classifiers as part of our participation in ACL 2019 Social Media Mining for Health Applications (SMM4H) shared task 1. The best-performed model on the test sets were trained on a merged corpus consisting of the datasets released by SMM4H 2017 and 2019. By using VUE, the corpus was randomly under-sampled with 2:1 ratio between the negative and positive classes to create an ensemble using the linear kernel trained with features including bag-of-word, domain knowledge, negation and word embedding. The best performing model achieved an F-measure of 0.551 which is about 5% higher than the average F-scores of 16 teams.

1   Introduction

Our team participated in the Social Media Mining for Health Applications (SMM4H) shared Task 1, which focus on the task of automatic classification of adverse effects mentions in tweets to distinguish tweets mentioned adverse effect (AE) from others(Weissenbacher et al., 2019).

2   Methods

AEC (Adverse Effect Classification) task is a typical classification problem. We used support vector machine (SVM) with the linear kernel to develop our classifiers. The training and validation sets released by the SMM4H 2019 organizers include 24,861 and 5,000 tweets, respectively. The organizers provided the entire training set but the validation set was downloaded by ourselves using the Twitter API. Unfortunately, only 2,887 tweets in the validation set can be downloaded from the Twitter website. In addition, we included the corpus released in AMIA-SMM4H 2017 for the same purpose, which contains 11,564 tweets (SarkerandGonzalez-Hernandez, 2017). We merged the two datasets and filtered out duplicate tweets to create a merged corpus for our model. The merged corpus contains 3,423 positive tweets and 31,858 negative tweets.

The imbalance ratio for the compiled corpus is 9.3, which is highly imbalanced. In order to develop classifiers with reliable performance, we implemented a vote-based under-sampling ensemble (VUE) technique Wang et al. (2018). VUE exploits all training examples in majority (negative) cases with under-sampling for creating an ensemble of SVM classifiers. It samples several subsets from the negative tweets without replacement and then create an ensemble by using each subset along with the minority cases (positive). The prediction can be determined by taking a majority vote among the separately created classifiers.

Proceedings of the 4th Social Media Mining for Health Applications (#SMM4H) Workshop & Shared Task, pages 117–119
Florence, Italy, August 2, 2019. ©2019 Association for Computational Linguistics

In order to extract features for training our classifiers, we first pre-processed tweets to replace URLs, dosages and Twitter specific characters with the corresponding symbols, and modified the numeral parts in each token to one as proposed in our previous work Dai et al. (2016). The preprocessed tweet was then processed by a tweet tokenizer (Owoputi et al., 2013) to generate tokens. Follow by the above step, each token was processed by Hunspell to detect spelling errors. If a token is considered to be misspelled, the first recommended correction is included as an alternative term for the token. Finally, we lowercased all tokens and used the Snowball stemmer(Porter, 2001) to perform stemming without removing any stop words.

After the above steps, we extracted the following features to train our SVM models:

- Bag-of-word features: we extracted unigram and bigram with TF-IDF (Term Frequency-Inverse Document Frequency) as the weighting scheme.

- Domain knowledge features: The presence of adverse drug reaction (ADR) or drug mentions were engineered as two binary features with the value of either 0 or 1. The occurrences of ADR and drug names were recognized by using the ADR mention recognizer developed in our previous work Dai et al. (2016) and Wang et al. (2018).

- Negation features: The feature set uses three flags to indicate the occurrence of an ADR mention is missing, positive or negated. If a tweet contains ADRs, the NegEx algorithm (Chapman et al., 2001) is employed to determine whether the occurrence is negated.

- Word embedding features: The word embedding features proposed in our previous work Wang et al. (2018) was developed. The features were generated by taking the mean across all tokens' embedding represented as a 400-dimensional vector based on the pre-trained tweet WE model released by (Godin et al., 2015).

3 Results

Figure 1 show the results of the 10-fold cross validation (CV) on the training set of the AEC task. The standard precision (P), recall (R) and F-measure (F) are used to report the performance. Configuration 1 is the VUE model trained with the developed features. After submitting the results, we developed configuration 2, which was a baseline model with the same features but didn't apply any imbalanced techniques. The above two

configurations of the developed classifiers were trained on the following three corpora:

1. SMM4H 2017 corpus
2. SMM4H 2019 corpus
3. SMM4H 2017+SMM4H 2019 corpus

During the participating the AEC task, we used configuration 1 with the above three corpora to conduct ablation experiments and submitted three runs corresponding to the first three configurations shown in Figure 1. The experimental results show that the VUE method has better recall but lower precision. The F-scores of VUE are better than baseline on the first two corpora. It is interesting to see that the baseline configuration performs better than VUE on the merged corpus.

4 Conclusion

In this paper, we briefly describe our systems developed for the SMM4H 2019 AEC task. Our best submitted run was based on the VUE model trained on a merged corpus. However, we noticed that by using the merged corpus, the baseline model which didn't exploit imbalanced technique performs better than that of VUE on the 10 fold CV. We will conduct error analysis to investigate the interesting results and compare the performance of other advanced imbalance techniques developed in our previous work Dai and Wang (2019).

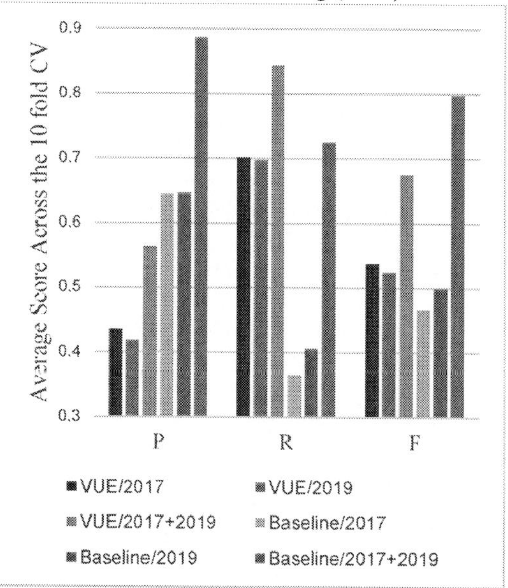

Figure 1: 10 fold CV on the training set of the AEC task.

References

Chapman, Wendy W, Bridewell, Will, Hanbury, Paul, Cooper, Gregory F, & Buchanan, Bruce G. (2001). A simple algorithm for identifying negated findings and diseases in discharge summaries. *Journal of biomedical informatics, 34*(5), 301-310.

Dai, Hong-Jie, Touray, Musa, Jonnagaddala, Jitendra, & Syed-Abdul, Shabbir. (2016). Feature engineering for recognizing adverse drug reactions from twitter posts. *Information, 7*(2), 27.

Dai, Hong-Jie, & Wang, Chen-Kai. (2019). Classifying Adverse Drug Reactions from Imbalanced Twitter Data. *International Journal of Medical Informatics.*

Godin, Fréderic, Vandersmissen, Baptist, De Neve, Wesley, & Van de Walle, Rik. (2015). *Multimedia Lab $@ $ ACL WNUT NER Shared Task: Named Entity Recognition for Twitter Microposts using Distributed Word Representations.* Paper presented at the Proceedings of the Workshop on Noisy User-generated Text.

Owoputi, Olutobi, O'Connor, Brendan, Dyer, Chris, Gimpel, Kevin, Schneider, Nathan, & Smith, Noah A. (2013). *Improved part-of-speech tagging for online conversational text with word clusters.* Paper presented at the Proceedings of the 2013 conference of the North American chapter of the association for computational linguistics: human language technologies.

Porter, Martin F. (2001). Snowball: A language for stemming algorithms. In.

Sarker, Abeed, & Gonzalez-Hernandez, Graciela. (2017). Overview of the Second Social Media Mining for Health (SMM4H) shared tasks at AMIA 2017. *Training, 1*(10,822), 1239.

Wang, Chen-Kai, Dai, Hong-Jie, Wang, Feng-Duo, & Su, Emily Chia-Yu. (2018). *Adverse Drug* Reaction *Post Classification with Imbalanced Classification Techniques.* Paper presented at the 2018 Conference on Technologies and Applications of Artificial Intelligence (TAAI).

Weissenbacher, Davy , Sarker, Abeed, Magge, Arjun, Daughton, Ashlynn, O'Connor, Karen, Paul, Michael, & Graciela, Gonzalez-Hernandez. (2019). *Overview of the Fourth Social Media Mining for Health (SMM4H) Shared Task at ACL 2019.* Paper presented at the Proceedings of the 2019 ACL Workshop SMM4H: The 4th Social Media Mining for Health Applications Workshop & Shared Task.

Detection of Adverse Drug Reaction mentions in tweets using ELMo

Sarah Sarabadani
Klick Health, Toronto, Canada.
ssarabadani@klick.com

Abstract

This paper describes the models used by our team in SMM4H 2019 shared task (Weissenbacher et al., 2019). We submitted results for subtasks 1 and 2. For task 1 which aims to detect tweets with Adverse Drug Reaction (ADR) mentions we used ELMo embeddings which is a deep contextualized word representation able to capture both syntactic and semantic characteristics. For task 2, which focuses on extraction of ADR mentions, first the same architecture as task 1 was used to identify whether or not a tweet contains ADR. Then, for tweets positively classified as mentioning ADR, the relevant text span was identified by similarity matching with 3 different lexicon sets.

1 Introduction and task description

Twitter is an ever-growing store of daily generated data. Given the huge number of tweets talking about drug-related issues, social media mining is applicable to areas such as pharmacovigilance (Lee et al., 2017; Nikfarjam et al., 2015; Ginn et al., 2014; Freifeld et al., 2014; Bian et al., 2012).

Tasks 1 and 2 focuses on detecting tweets with ADR and identifying location of mentions. We are provided with 25,672 tweets (2,374 positive and 23,298 negative) and approximately 5,000 unlabeled tweets as a validation set. For the second task, a subset of 2,367 tweets from the first task was provided (1,212 positive and 1,155 negative). The evaluation data comprises 1,000 tweets (~500 positive, ~500 negative).

2 Preprocessing

Stop words and punctuations were removed from tweets and all drug names found in the FDA's Approved Drug Products list[1] were replaced by the word "drug". Word stemming and tokenization were performed using nltk python library.

3 Methods

3.1 task 1

For this task, we used 4 deep learning models. The architecture of the first 3 models were relatively similar, differing in the embedding layer.

The first model involves character embedding with dimension equal to the total number of unique characters in training set including emojis. The output of this layer is fed to a series of 6 convolutional neural network layers (CNNs) with ReLU activation. Each CNN used 256 filters, with a filter size of 7 for the first two layers and 3 for the rest. Max pooling with size 3 was used for the first two and last CNNs. The CNNs' output was fed into a bidirectional LSTM (Bi-LSTM) with 2*200 units, whose output was flattened to feed into two dense layers. We used two fully connected layers with 1024 units each, ReLU activation, and dropout of 0.5. Finally, we used a dense layer with size two and softmax activation. We used Adam as the optimizer and binary cross-entropy as the loss function. The model was trained with 10 epochs and batch size of 128.

The second architecture was identical to the first, except the first layer was a word embedding using GloVe[2] pre-trained on Twitter data with embedding dimension of 100.

---

[1] https://www.fda.gov/drugs/drug-approvals-and-databases/drugsfda-data-files

[2] https://nlp.stanford.edu/projects/glove/

Proceedings of the 4th Social Media Mining for Health Applications (#SMM4H) Workshop & Shared Task, pages 120–122
Florence, Italy, August 2, 2019. ©2019 Association for Computational Linguistics

The third model was a concatenation of word and character embeddings. We combined the Bi-LSTM output of the first and second models and then applied dense layers as before.

After building the above models, we tried to improve the outcomes by adding layers and features. We used a multi-head self-attention with an attention width of 15 and ReLU activation. We also explored the effect of sentiment features. Since the data classes were imbalanced, we tried to make class sizes equal by downsampling and upsampling. In downsampling, samples from the majority class (tweets without ADR mentions) were randomly sampled without replacement. In upsampling we did the opposite, adding samples from the minority class with replacement. None of these strategies substantially altered our baseline results.

In our final model, we used ELMo (Peters et al., 2018) (Embeddings from Language Models) with 1024 dimensions. In contrast to traditional word embeddings such as GloVe and word2vec, ELMo assigns each word to a vector as a function of the entire sentence containing that word. Therefore, the same word can have different embeddings depending on its context. Since ELMo already captures character-level information under the hood, we decided to encircle the complexity inside the embedding layer and used only two additional dense layers with 256 and 2 units, using ReLU and softmax activations, respectively.

3.2 Methods for task 2

To identify the text spans of ADR mentions, first the model developed for task 1 was used to determine whether each tweet mentions an ADR. Then the similarity between each tweet and 3 different lexicon sets (Nikfarjam et al.[3], MedDRA (Medical Dictionary for Regulatory Activities)[4], and CHV (Consumer Health Vocabulary) [5]) was measured.

To calculate similarity, each tweet and lexicon was converted to a set of word stems. Since similarity measures such as cosine or Jaccard are highly affected by other non-ADR words, we defined similarity as the percent of word stems of a lexicon that exist in a tweet. For each tweet, only lexicons with a 100% match were kept.

4   Results, discussion, and next steps

Among all architectures, the best results came from ELMo embedding (F1 = 0.64). Therefore, we only submitted ELMo results with 5, 10, and 15 epochs. The model performed less well for the validation set (F1 = 0.41), below the average F1 score of 0.50 among all teams, which might result from overfitting. Using more sophisticated architecture after the embedding layer might improve performance.

Since task 2's performance depends strongly on task 1, we also scored lower on this task compared to the team average (0.40 vs. 0.54). Since ADR phrases and tweets do not always lexically match, approaches such as named entity recognition (NER) might perform better.

Other approaches to improve performance:
Task 1:
- Try other embeddings such as BERT

- Experiment with more complex architectures after the ELMo layer

- Add part of speech (POS) tags

- Add topic modeling and tweet cluster features

Task 2:
- Search Twitter for keywords from lexicon sets to augment the training set with new tweets which mention ADRs

- Try NER

Acknowledgment

I would like to thank Maheedhar Kolla who provided insight and expertise that significantly assisted this work.

I would also like to show my gratitude to Peter Leimbigler for comments that greatly improved the manuscript.

Finally, special thanks go to Alfred Whitehead for supporting me to participate in this challenge.

---

[3] http://diego.asu.edu/Publications/ADRMine.html
[4] https://www.meddra.org/how-to-use/support-documentation/english

[5] https://www.nlm.nih.gov/research/umls/sourcereleasedocs/current/CHV/

References

Jiang Bian, Umit Topaloglu, and Fan Yu. (2012, October). Towards large-scale twitter mining for drug-related adverse events. In *Proceedings of the 2012 international workshop on Smart health and wellbeing* (pp. 25-32). ACM.

Clark C. Freifeld, John S. Brownstein, Christopher M. Menone, Wenjie Bao, Ross Filice, Taha Kass-Hout, and Nabarun Dasgupta. (2014). Digital drug safety surveillance: monitoring pharmaceutical products in twitter. *Drug safety, 37*(5), 343-350.

Rachel Ginn, Pranoti Pimpalkhute, Azadeh Nikfarjam, Apurv Patki, Karen O'Connor, Abeed Sarker, Karen Smith, and Graciela Gonzalez. (2014, May). Mining Twitter for adverse drug reaction mentions: a corpus and classification benchmark. In *Proceedings of the fourth workshop on building and evaluating resources for health and biomedical text processing* (pp. 1-8).

Kathy Lee, Ashequl Qadir, Sadid A. Hasan, Vivek Datla, Aaditya Prakash, Joey Liu, and Oladimeji Farri. (2017, April). Adverse drug event detection in tweets with semi-supervised convolutional neural networks. In *Proceedings of the 26th International Conference on World Wide Web* (pp. 705-714). International World Wide Web Conferences Steering Committee.

Azadeh Nikfarjam, Abeed Sarker, Karen O'connor, Rachel Ginn, and Graciela Gonzalez. (2015). Pharmacovigilance from social media: mining adverse drug reaction mentions using sequence labeling with word embedding cluster features. *Journal of the American Medical Informatics Association, 22*(3), 671-681.

Matthew E. Peters, Mark Neumann, Mohit Iyyer, Matt Gardner, Christopher Clark, Kenton Lee, and Luke Zettlemoyer. (2018). Deep contextualized word representations. arXiv preprint arXiv:1802.05365.

Davy Weissenbacher, Abeed Sarker, Arjun Magge, Ashlynn Daughton, Karen O'Connor, Michael Paul, Graciela Gonzalez-Hernandez. Overview of the Fourth Social Media Mining for Health (SMM4H) Shared Task at ACL 2019. *In Proceedings of the 2019 ACL Workshop SMM4H: The 4th Social Media Mining for Health Applications Workshop & Shared Task*

Adverse drug effect and personalized health mentions
CLaC at SMM4H 2019, Tasks 1 and 4

Parsa Bagherzadeh, Nadia Sheikh, Sabine Bergler
CLaC Labs, Concordia University
Montreal, Canada
{ p_bagher, n_she, bergler } @encs.concordia.ca

Abstract

CLaC labs participated in Task 1 and 4 of SMM4H 2019. We pursed two main objectives in our submission. First we tried to use some textual features in a deep net framework, and second, the potential use of more than one word embedding was tested. The results seem positively affected by the proposed architectures.

1 Introduction

The ongoing SMM4H challenge tasks define evolving challenges defined on Twitter data (Weissenbacher et al., 2019). The intention of epidemiologists is to detect mentions of health issues early on Twitter. One of the challenges is to detect real reports of personally experienced health issues and to distinguish them from generalizations, hypotheticals, news, and institutional advice.

Task 1 of SMM4H 2019, "Automatic classification of adverse effects mentions in tweets", asks to distinguish tweets that report an adverse drug effect (AE) from those that do not. Training data consists of 25,672 tweets with imbalanced distribution: 2,374 positive and 23,298 negative labels. An example of an adverse effect mention in a tweet is:

> *saphris gives me a mad appetite omg i
> hate this*

Task 4 is on "Generalizable identification of personal health experience mentions". Two specialized training sets were released , "flu vaccination" and "flu infection", comprising approximately 6,200 and 1,100 tweets. Task 4 training data was balanced. A sample positive tweet from this task is:

> *I must say that flu shot packed a punch.
> #WorstInoculationEver*

The CLaC submission to SMM4H 2019 had three general goals: first, to experiment with architectures that can address both tasks, second, to compare different word embeddings for their individual, but also their combined effectiveness, and third, to test whether we can augment the basic word vectors input with additional local and global knowledge from word lists and text preprocessing. The experiments remain inconclusive, due to an error in our submission pipeline.

2 Word embeddings

We experimentd with three types of word embeddings: BERT (a Transformer-based Bidirectional representation) (Devlin et al., 2018) (BERT-Base, Uncased)[1]; Word2Vec (Mikolov et al., 2013) trained on Sentiment140 [2] as well as training data from SMM4H 2018 and 2019 (all tasks) using Gensim package (Řehůřek and Sojka, 2010); and Glove word embeddings, pretrained on tweets (Pennington et al., 2014).

3 Textual features

Use of textual features as external source of knowledge has recently been the topic of interest (Sennrich and Haddow, 2016), (Ebert et al., 2015). We preprocess the tweets using the ANNIE Twitter Tokenizer (Cunningham et al., 2002), the Hashtag Tokenizer (Maynard and Greenwood, 2014), and the Stanford Part-Of-Speech Tagger with a model trained on tweets (Toutanova et al., 2003). We determine negation and modality spans using (Rosenberg et al., 2012). We use the Diego Lab ADR wordlist (Nikfarjam et al., 2015) to annotate terms appropriate for negative effects and health concerns.

---

[1] https://github.com/google-research/bert
[2] http://help.sentiment140.com/for-students

Proceedings of the 4th Social Media Mining for Health Applications (#SMM4H) Workshop & Shared Task, pages 123–126
Florence, Italy, August 2, 2019. ©2019 Association for Computational Linguistics

User mentions (@) were removed from the tweets. URLs are annotated, as are the first person personal pronouns *I, my, mine.*

Negation and modality The span of negation and modality is determined using (Rosenberg et al., 2012) and projected onto the token representation: tokens present in the span of a negation or modality are indicated by a binary flag appended to the respective word vector (see Figure 1). The presence of negation and/or modality might reflect uncertainty in a given tweet and it may not convey facts.

URL Tweets about a personal experience do not usually include a URL. Specifically for Task 4, 80% of the tweets including a URL are negative. A binary URL feature encodes presence or absence of a URL in the tweet.

POS embedding We experimented with the notion of part of speech embeddings to address sparsity. Here, a representation for each POS tag is obtained using Word2vec by training on a POS tagged corpus (instead of words themselves). We use the Penn tree bank tag set (36 tags) with a window size of 5.

ADR lexicon Terms from the Diego Lab adverse drug reaction lexicon (Nikfarjam et al., 2015) are indicated as a binary, tweet level feature, in order to increase recall.

First person personal pronoun First person pronouns *I, my,* and *mine* are indicated at token level by a separate binary feature. In both tasks, a personal experience is more likely to be a positive sample, therefore, enhancing recall.

	I	should	n't	have	gotten	that	flu	shot
W2V	...	...	...		...	...	...	...
Neg.	0	0	0	1	1	1	1	1
Mod.	0	0	1	1	1	1	1	1
1st	1	0	0	0	0	0	0	0

Figure 1: Feature vector encoding

3.1 System architecture

Our system has two parallel branches and is trained in two stages. One branch works only with BERT word embeddings, the other branch works on our concatenated token level features plus word embeddings (Word2Vec/Glove) shown in Figure 1. The input vectors of each branch are fed into Bi-LSTMs and are followed by attention and finally two softmax decision neurons.

After optimizing each branch with binary cross-entropy loss, the parameters of the networks are frozen for the second stage of training. We train an SVM on the input vector that concatenates class probabilities provided by the softmax neurons with the tweet level features, ADR and URL.

The network is optimized using the Adam optimizer (Kingma and Ba, 2014) with learning rate $lr = 0.001$ for 5 epochs (for both tasks). For Task 1, the class weights of $cw_{pos} = 1$ and $cw_{neg} = 0.4$ are used as thresholds for positive and negative samples respectively. For the SVM, the RBF kernel is used with $\gamma = 0.001$. The hyper-parameters have been chosen by cross validations. The first stage deep net learning is implemented using Keras [3] and the second stage SVM classification is implemented using Scikit-learn (Pedregosa et al., 2011).

4 Development phase

During the development phase we considered a number of different features and performed an ablation study with more than 130 different configurations. For this phase, 22,000 and 3,672 samples were considered for training and test sets respectively.

An interesting observation was the different behavior of word embeddings in the presence of language features. For Task 1, Glove embeddings usually performed higher, whereas in Task 4, Word2Vec embeddings were generally superior. In Task 1, adding textual features to Word2Vec embeddings resulted in a decrease in performance, however, adding the same features to Glove re-

---

[3] https://keras.io

Table 1: Development results for Task 1. Submitted configurations are indicated by *

	Prec.	Rec.	F1
Glove	0.41	**0.73**	0.52
BERT	0.56	0.50	0.53
Glove+ADR	0.46	0.67	0.55
Glove+BERT	0.49	0.64	0.55
Glove+Mod+BERT	0.53	0.57	0.55
Glove+Neg+BERT	0.48	0.61	0.54
Glove+Neg+Mod+BERT	0.58	0.55	0.56
Glove+BERT+ADR	0.53	0.64	0.58
Glove+Neg+Mod+ADR	0.49	0.65	0.56
* Glove+Neg+Mod+ADR+BERT	0.54	0.64	**0.59**
W2V	0.42	0.65	0.51
W2V+ADR	0.39	0.67	0.49
* W2V+BERT	**0.59**	0.53	0.56
W2V+1st	0.48	0.62	0.54
* W2V+1st+BERT	0.52	0.63	0.57

sulted in increased performance. This effect was small, but persistent across ablation of the other features, and we concluded that the different behaviors of the embedding vectors could be leveraged in an ensemble situation.

For Task 1, the ADR word list generally increased recall in our ablation studies, demonstrating that domain specific gazetteer lists can effectively supplement training data. In combination with Glove, textual features such as negation and modality increased precision, but diminished recall. Adding ADR to this combination (*Glove+Neg+Mod+BERT*) compensates for the drop in recall without significantly decreasing precision. The results also corroborates the hypothesis that the *1st* feature enhances the recall (*W2V+1st* and *W2V+1st+BERT* compared to *W2V* and *W2V+BERT*).

Looking at the confusion matrix reveals that the model (specifically *Glove+BERT*) associates drug mentions in the subject position with positive labels, incurring a considerable amount of false positives, see for instance:

this lozenge has my sore throat fading

paxil makes you susceptible to sunburns?

The ADR feature (*Glove+ADR+BERT*) reduces these false positives while it causes other instances of false positives. As mentioned before, ADR generally increases recall, but in some configurations with Glove it has increased precision which is interesting and we will study it in more detail.

Modality reduces false positives and is the most effective token level textual feature. Two instances of false positives (in *Glove+BERT*) which are correctly classified in the presence of modality are:

Table 2: Development results for Task 4. Submitted configurations are indicated by *

F1	Prec.	Rec.	
W2V	0.70	**0.88**	0.78
BERT	0.78	0.82	0.80
W2V+BERT	0.76	0.85	0.80
W2V+Mod	0.72	0.87	0.79
W2V+POS	0.76	0.81	0.79
W2V+URL	0.76	0.84	0.80
* W2V+URL+BERT	**0.83**	0.79	0.81
W2V+1st+URL	0.77	0.83	0.80
* W2V+1st+URL+BERT	0.81	0.81	0.81
W2V+Mod+POS+URL	0.78	0.85	0.81
* W2V+Mod+POS+URL+BERT	0.81	0.84	**0.83**

*i'm sucha psycho when i study already if i ever took adderall i **would** probably explode*

*seroquel **can** have potential fatal effects when taken & being in direct sunlight for extended periods. can i get you a bottle a tanning bed?*

When combined with Glove, we observed that the negation feature degrades the F1 score, however, it inter-plays well with the modality feature.

For Task 4, combining textual features with Word2Vec increases precision. The URL feature by itself increases precision even more, but incurs a larger drop in recall.

5 Evaluation phase

Task 1 We submitted three configurations to Task 1: Glove with our textual features, W2V alone, and W2V with the first person pronoun feature (all used in an ensemble with BERT). These were not our top performing configurations during development, rather we included W2V to bridge to Task 4 and we included two runs with different textual features and one without. The performance of our system in the competition is provided in Tables 3, the competition performance of all three models is commensurate with our development results with $\pm 2\%$ in F1 measure. Moreover, the three configurations performed near identically and all three were above the competition mean.

It is interesting to note that the Word2Vec embeddings trained on Sentiment140 data proved as effective on this data set as Glove with the textual features, in contrast to our development experiments. We interpret the fact that W2V in an ensemble with BERT lies above the competition's mean to confirm the importance of our genre selection for Word2vec training.

Table 3: CLaC competition results for Task 1

	Prec.	Rec.	F1
W2V+ BERT	0.54	0.60	0.57
Glove+Neg+Mod+ ADR+BERT	0.52	0.60	0.56
W2V+ 1st+BERT	0.51	0.59	0.55
Competition mean	0.53	0.50	0.50

Task 4 Our three submissions for Task 4 were all based on Word2vec and the URL feature. Results, however, diverge drastically from our development runs, where runs scored between 75-85%

Table 4: CLaC competition results for Task 4

	Prec.	Rec.	F1
Condition 1			
W2V+Mod+POS+URL+BERT	0.84	0.32	0.47
W2V+1st+URL+BERT	0.83	0.42	0.56
W2V+BERT+URL	0.75	0.29	0.42
Condition 2			
W2V+Mod+POS+URL+BERT	0.42	0.19	0.26
W2V+1st+URL+BERT	0.44	0.12	0.20
W2V+BERT+URL	0.44	0.12	0.20
Condition 3			
W2V+Mod+POS+URL+BERT	0.71	0.26	0.38
W2V+1st+URL+BERT	0.62	0.26	0.37
W2V+BERT+URL	0.62	0.26	0.37
Overall			
W2V+Mod+POS+URL+BERT	0.70	0.28	0.40
W2V+1st+URL+BERT	0.75	0.29	0.42
W2V+BERT+URL	0.74	0.33	0.46
Competition mean	0.90	0.58	0.70

F1 measure. The official results in Table 4 demonstrate.

6 Conclusions

We participated in the SMM4H 2019 shared task with two major ideas. First, we tried to use textual annotations in a deep net architecture and specifically proposed encodings for negation, modality, and use of a gazetteer list. Our observations during the development phase showed that textual features are effective for enhancing the performance of the system but that standard embedding vectors without additional textual features give comparable performance on these datasets.

Our second idea was to have more than one type of embedding in our system to have an ensemble and try to aggregate the predictions using a support vector machine rather than using a simple majority voting. This worked well, but again, on the datasets of this challenge, the computational overhead seems questionable for the degree of improvement achieved.

References

H Cunningham, D Maynard, K Bontcheva, and V Tablan. 2002. GATE: A framework and graphical development environment for robust NLP tools and applications. In *Proceedings of the 40th Anniversary Meeting of the Association for Computational Linguistics (ACL'02)*.

J Devlin, M-W Chang, K Lee, and K Toutanova. 2018. BERT: Pre-training of deep bidirectional transformers for language understanding. ArXiv:1810.04805v1 [cs.CL].

Sebastian Ebert, Ngoc Thang Vu, and Hinrich Schütze. 2015. A linguistically informed convolutional neural network. In *Proceedings of the 6th workshop on computational approaches to subjectivity, sentiment and social media analysis*, pages 109–114.

DP Kingma and J Ba. 2014. Adam: A method for stochastic optimization. ArXiv:1412.6980 [cs.LG].

DG Maynard and MA Greenwood. 2014. Who cares about sarcastic tweets? Investigating the impact of sarcasm on sentiment analysis. In *LREC 2014 Proceedings*.

T Mikolov, I Sutskever, K Chen, GS Corrado, and J Dean. 2013. Distributed representations of words and phrases and their compositionality. In *Advances in Neural Information Processing Systems*.

A Nikfarjam, A Sarker, K Oconnor, R Ginn Rachel, and G Gonzalez. 2015. Pharmacovigilance from social media: mining adverse drug reaction mentions using sequence labeling with word embedding cluster features. *Journal of the American Medical Informatics Association*, 22(3):671–681.

F Pedregosa, G Varoquaux, A Gramfort, V Michel, B Thirion, O Grisel, M Blondel, P Prettenhofer, R Weiss, V Dubourg, J Vanderplas, A Passos, D Cournapeau, M Brucher, M Perrot, and E Duchesnay. 2011. Scikit-learn: Machine learning in Python. *Journal of Machine Learning Research*, (12).

J Pennington, R Socher, and C. D. Manning. 2014. Glove: Global vectors for word representation. In *Empirical Methods in Natural Language Processing (EMNLP)*, pages 1532–1543.

R Řehůřek and P Sojka. 2010. Software framework for topic modelling with large corpora. In *Proceedings of the LREC 2010 Workshop on New Challenges for NLP Frameworks*.

S Rosenberg, H Kilicoglu, and S Bergler. 2012. Clac labs: Processing modality and negation. *Working Notes for QA4MRE Pilot Task at CLEF*.

Rico Sennrich and Barry Haddow. 2016. Linguistic input features improve neural machine translation. *arXiv preprint arXiv:1606.02892*.

K Toutanova, D Klein, ChD Manning, and Y Singer. 2003. Feature-rich part-of-speech tagging with a cyclic dependency network. In *Proceedings of NAACL-HLT*.

D Weissenbacher, A Sarker, A Magge, A Daughton, MJ Paul, and G Gonzalez-Hernandez. 2019. Overview of the fourth social media mining for health (SMM4H) shared tasks at ACL 2019. In *Proceedings of the 2019 ACL Workshop SMM4H: The 4th Social Media Mining for Health Applications Workshop and Shared Task*.

MIDAS@SMM4H-2019: Identifying Adverse Drug Reactions and Personal Health Experience Mentions from Twitter

Sarthak Anand,[3] Debanjan Mahata,[1] Haimin Zhang,[1] Simra Shahid,[2]
Laiba Mehnaz,[2] Yaman Kumar,[5] Rajiv Ratn Shah,[4]

[1]Bloomberg, USA, [2]DTU-Delhi, India, [3]NSIT-Delhi, India, [4]IIIT-Delhi, India, [5]Adobe, India,
sarthaka.ic@nsit.net.in, dmahata@bloomberg.net, hzhang449@bloomberg.net,
simrashahid_bt2k16@dtu.ac.in, laibamehnaz@dtu.ac.in, ykumar@adobe.com,
rajivratn@iiitd.ac.in

Abstract

In this paper, we present our approach and the system description for the Social Media Mining for Health Applications (SMM4H) Shared Task 1,2 and 4 (2019). Our main contribution is to show the effectiveness of Transfer Learning approaches like BERT and ULM-FiT, and how they generalize for the classification tasks like *identification of adverse drug reaction mentions* and *reporting of personal health problems* in tweets. We show the use of stacked embeddings combined with BLSTM+CRF tagger for identifying spans mentioning adverse drug reactions in tweets. We also show that these approaches perform well even with imbalanced dataset in comparison to undersampling and oversampling.

1 Introduction

Drugs administered for alleviating common sufferings are the fourth biggest cause of death in US, following cancer and heart diseases (Giacomini et al., 2007), making it one of the most important medical problems for the human society. While heart diseases and cancer are commonly reported and studied, adverse reactions to drugs either goes unreported or is confused or lost within other narratives. While it is the onus of the government and the society as a whole to tackle the first task, the second one is an overwhelmingly computational task.

With the advent of universal internet and smartphones, reportrage of incidents is generally increasing, thanks to a host of social media platforms like *Twitter, Facebook, Instagram, etc.* Hence, this unique situation presents a challenging as well as rewarding opportunity to improve our current computational systems for dealing with the existing incidents more sensibly and increase their reportage with the use of electronic media.

With this motivation, four shared tasks were conducted as part of *Social Media Mining for Health Applications (SMM4H) Workshop 2019* (Weissenbacher et al., 2019). Our team participated in Tasks - 1, 2 and 4 of the workshop. The problems for these tasks were:

Problem Definition Sub-task 1: *Given a labeled dataset D of tweets, the objective of the task is to learn a classification/prediction function that can predict a label l for a given tweet t, where l ∈ {reporting adverse effects of drugs (ADR) - 1, no adverse effects of drugs (non-ADR) - 0}.*

Example of tweets mentioning adverse drug reactions:
- *I feel siiiiiiiiiiiiiiick. Damn you venlafaxine.*
- *Who need alcohol when you have gabapentin and tramadol that makes you feel drunk at 12oclock.*

Problem Definition Sub-task 2: The motive of this sub-task is to first discern ADR tweets from the non-ADR ones and then identify the span of a tweet where an adverse drug effect is reported.

An example of a span from a tweet that represents the mention of adverse drug reactions:
- *losing it. could not remember the word power strip. wonder which drug is doing this memory lapse thing. my guess the cymbalta. #helps*, where *not remember* is the adverse drug reaction that needs to be identified and extracted from the tweet, which is most likely caused by the intake of the drug named *cymbalta*.

Problem Definition Sub-task 4: *Given a labeled dataset D of tweets, the objective of the task is to learn a classification/prediction function that can predict a label l for a given tweet t, where l ∈ {reporting personal health experience - 1, no mention of personal health experience - 0}.*

Example of tweets reporting personal health experience mentions:
- *This flu shot got my arm killing me.*

Proceedings of the 4th Social Media Mining for Health Applications (#SMM4H) Workshop & Shared Task, pages 127–132
Florence, Italy, August 2, 2019. ©2019 Association for Computational Linguistics

• *man i am so sick i feel terrible i got all the symptoms of the swine flu i am scared.*

Our Contributions: Towards the objectives of the tasks as described above, we present some of our contributions in this paper:

1. We train ULMFit and BERT models for Tasks 1 and 4, and show that these models are agnostic to the effects of undersampling and oversampling, given a highly imbalanced dataset.

2. We make an initial attempt in studying the effectiveness of transfer learning using ULMFit and BERT for the problems in the domain of health care pertaining to the shared tasks.

3. We show the use of stacked embeddings combined with BLSTM+CRF tagger for identifying spans mentioning adverse drug reactions in tweets.

4. We also show the use of combining pretrained BERT embeddings with Glove embeddings fed to a BLSTM text classifier for sub-task-1 and sub-task-4.

2 Related Work

In general, self reporting of drug effects by patients is a highly noisy source of data. However, even after being noisy, it captures quite a lot of information which might not be available in other cleaner sources of data such as limited clinical trials or a doctor's office (Leaman et al., 2010). Taking cognizance of this, the International Society of Drug Bulletins in 2005 said, "...patient reporting systems should periodically sample the scattered drug experiences patients reported on the internet...". This is an upcoming branch which lies at the intersection of information systems and medicine - *pharmacovigilance* (Leaman et al., 2010). Detecting and tracking information about certain diseases has been the focus of quite a lot of work (Nakhasi et al., 2012; Paul and Dredze, 2011). For instance, cancer investigation (Ofran et al., 2012), flu (Aramaki et al., 2011; Lamb et al., 2013) and depression (De Choudhury et al., 2013; Yazdavar et al., 2017). There has been some work in the domain of pharmacovigilance (Mahata et al., 2018b,a,c; Mathur et al., 2018; Sarker et al., 2018), recently as well.

The body of works most relevant to ours is the one which uses transfer learning on health domain.

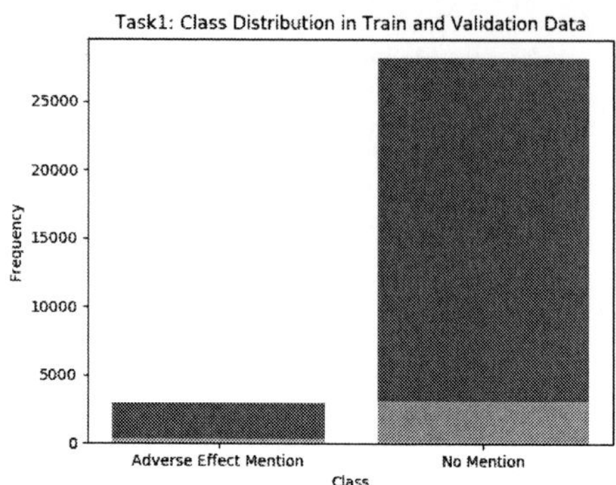

Figure 1: Distribution of classes in Train and Validation datasets for Sub-Task-1 (Identifying ADR and non-ADR tweets)

Normally, data in health domain is harder to get and process. Thus, many researchers have resorted to using transfer learning in order to deal with the data paucity. The works using transfer learning generally use word embeddings in order to improve the generalization of classification to unseen textual cases. In the context of this work we heavily use ULMFit (Howard and Ruder, 2018) and BERT (Devlin et al., 2018) for our experiments and make an initial attempt on how transfer learning in the domain of health works using them for the different text classification tasks of Social Media Mining for Health Workshop. Next, we give a brief description of the datasets used in this work for the different tasks.

3 Dataset

The dataset for the shared tasks was collected from the social networking website, *Twitter*. It consists of mentions of drug effects and other health related issues.

1. For the shared task 1, a total of 25,672 tweets are made available for training, out of which 2,374 contain adverse drug reaction (ADR) mention and the rest (23,298) do not. Only training data was provided by the organizers. For performing our experiments we segmented the provided dataset into train and validation splits. Figure 1 shows the distribution of data in the training and validation splits. The evaluation metric for this task was the F-score for the ADR class. Due to appreciable data bias, for the various experiments for this subtask, we oversample ADR tweets

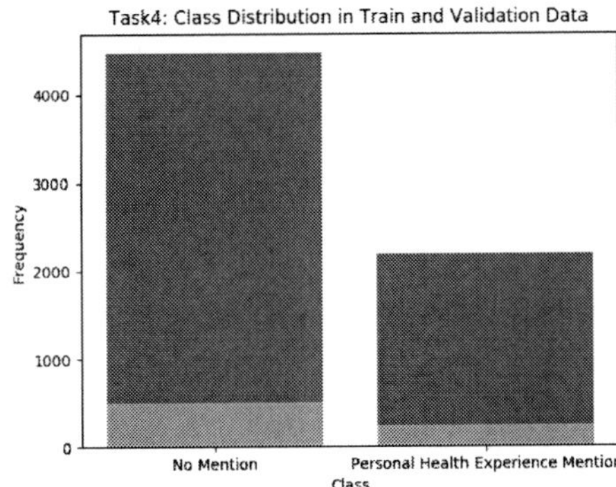

Figure 2: Distribution of classes in Train and Validation datasets for Sub-Task-4 (Identifying reporting of personal health experience mentions and no mentions in tweets)

and undersample non-ADR tweets. For oversampling, we just copy the ADR tweets and for undersampling, we randomly select a set of tweets such that the total number of tweets in both the sets becomes equal. For instance *"feeling a little dizzy from the quetiapine i just popped!"* represents a positive sample from the dataset while *"don't say no to pills! latuda won't kill!"* is a non-ADR tweet. We also try imbalanced proportions such as from 1:2 to 1:10 as well.

2. For the shared task 2, we got a total of 2,367 tweets out of which 1,212 were positive and 1,155 were negative. In the positive samples, the ADR portion was marked. For instance, the tweet *"friends! anybody taken #cipro? (antibiotic) complications?? big side effect is tendon rupture...figured my dr would know better?"* is an ADR tweet and the portion *"tendon rupture"* is where the author of the tweet mentions about ADR.

3. For the shared task 4, we were given a total of 10,876 tweets out of which only 7,388(67.9%) of the tweets were available on twitter for downloading. A total of 3,598 were positive and the rest were negative in original data. The positive tweets in this case contained a personal mention of ones health (for example, sharing health status or opinion) where as negative samples contained a generic discussion of the health issue, or some unrelated mention of the word. For instance, 9,832 is an example of tweet which

contains flu-vaccination context in original data. Similarly, in the tweet 1,046, the author tries to discuss disease context of flu. For the available data we had 2,426 positive combined and 4,962 negative samples where the author is initiating general health discussion as opposed to mentioning any particular context of flu. For performing our experiments we segmented the provided dataset into train and validation splits. Figure 2 shows the distribution of data in the training and validation splits.

3.1 Preprocessing

Before feeding the dataset to any machine learning model we took some steps to process the data. We point to those steps in this section. Normalization of tokens were done using some hand-crafted rules mainly for dealing with short forms such as *thru*(through), *abt*(about), etc. The '@*user*' and URL tokens were removed. The hashtags that contained two or more words were segmented into their component words using *ekphrasis* library[1]. For example *#NotFeelingWell* was converted to not feeling well.

3.2 Training Models

For all the tasks, we mainly concentrated in training recently introduced ULMFit and BERT models that are well known for their transfer learning capabilities and generalizing well for various natural language processing tasks across different domains. We describe our models in this section. We extensively used fast.ai[2], bert[3], and flair[4] for training our models related to all the tasks. The different models trained and their corresponding hyperparamaters chosen for the tasks are presented in Table 1. We provide their brief description next.

ULMFiT- We used ULMFit (Howard and Ruder, 2018) for tasks 1 and 4. One of the main advantages of training ULMFiT is that it works very well for a small dataset as provided in the task and also avoids the process of training a classification model from scratch. This avoids overfitting. We have used the base (*fast.ai*) implementation of this model.

The ULMFiT model has mainly two parts, the language model and the classification model.

---

[1] https://github.com/cbaziotis/ekphrasis
[2] http://nlp.fast.ai/category/classification.html
[3] https://github.com/google-research/bert
[4] https://github.com/zalandoresearch/flair

Tasks	Models	Hyperparameters
Task 1 (Identification of Tweets mentioining ADR)	BERT (Submission 1)	batch_size=32, learning_rate=2e-5, epochs=4
	ULMFit (Submission 2)	batch size=72, learning_rate= 3e-2, bptt=70, epochs= 8, embedding_size=400, hidden_size=1150, number_of_layers=3
	BLSTM (Submission 3)	Pretrained Embeddings - BERT + Twitter Glove learning_rate=0.1, mini_batch_size=32, anneal_factor=0.5, patience=5, max_epochs=50, lstm_units=512, dense-size=256
Task 2 (ADR span extraction from Tweets)	BLSTM + CRF (Submission 1)	Stacked Pretrained Embeddings - BERT+Twitter Glove hidden units=256, learning_rate=0.1, epochs=150, batch_size=32
	BLSTM + CRF (Submission 2)	Stacked Pretrained Embeddings - BERT+Twitter Glove + Flair hidden units=256, learning_rate=0.1, epochs=150, batch_size=32
Task 4 (Identification of Tweets reporting personal health experience)	BERT (Submission 1)	batch_size=32, learning_rate=2e-5, epochs=4
	BLSTM (Submission 2)	Pretrained Embeddings - BERT + Twitter Glove learning_rate=0.1, mini_batch_size=32, anneal_factor=0.5, patience=5, max_epochs=50, lstm_units=512, dense-size=256
	BLSTM (Submission 3)	Pretrained Embeddings - BERT + Flair learning_rate=0.1, mini_batch_size=32, anneal_factor=0.5, patience=5, max_epochs=50, lstm_units=512, dense-size=256

Table 1: Model architectures and their corresponding hyperparameters of all the submissions by team MIDAS for sub-tasks 1, 2 and 4.

We observe that fine-tuning the language model on a larger dataset provides a significant improvement in the performace (Tuhin Chakrabarty, 2019). Therefore, we fine-tune the language model over 1,90,823 tweets containing 250-drug related mentions (Sarker and Gonzalez, 2015). Default (*fast.ai*) parameters were used to train the language models. Finally, we find the best hyperparameters and train the classifier over the original training data.

BERT - We use the provided Tensorflow implementation of BERT and fine-tune BERT-base-uncased. We find the best parameters and train the model over original dataset.

BLSTM - We train a bidirectional LSTM text classifier and feed different types of pretrained embeddings as presented in the Table 1. It is important to note that due to the long time needed for training the BLSTM models with the embeddings and unavailability of GPUs, we could not finish the training before submitting our results for the test data provided by the organizers. We would like to make our predictions on the final model and keep it as a future work.

BLSTM+CRF Tagger - We treated the problem posed in sub-task 2 as a named entity extraction and recognition problem. The text span corre-sponding to an adverse drug reaction mention is treated as an entity, that further needs to be classified into one of the two categories ADR or non-ADR. Following the current state-of-the-art, we trained a BLSTM+CRF tagger implemented in the flair library (referred above). Apart from that, we also used the BLSTM+CRF architecture with two different combinations of stacked embeddings.

Next, we present the results obtained on the test data provided by the organizers for sub-tasks 1, 2 and 4.

4 Results

4.1 Task-1: Identifying Tweets Mentioning Adverse Drug Reactions

Model	F1	Precision	Recall
BERT	0.5759	0.5615	0.5911
ULMFiT	**0.5988**	0.6647	0.5447
BLSTM	0.5196	0.5891	0.4649

Table 2: Results for Task-1: Identifying Tweets Mentioning Adverse Drug Reactions

Table 2, presents the F1 scores on the test data for sub-task 1. The ULMFit model showed the best performance. As already mentioned, the data provided for sub-task 1 was highly imbal-

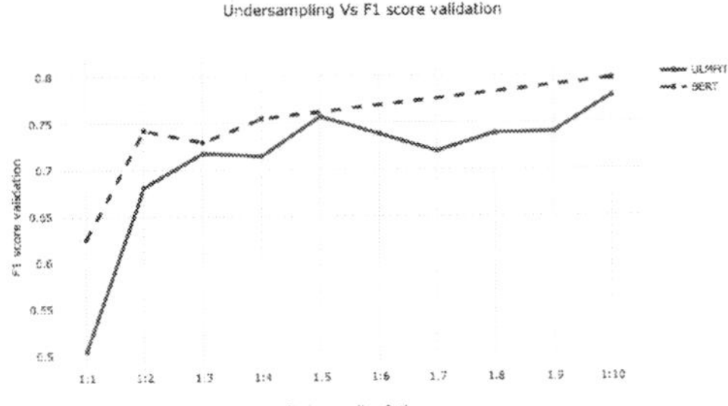

Figure 3: F1 score for ULMFit and BERT models trained on differently undersampled ratio of the classes (ADR : non-ADR).

anced. We performed undersampling with different ratios of the classes (ADR : non-ADR). Figure 3, presents the performance of ULMFit and BERT models on the training data for different undersampling ratios. We also tried oversampling, but didn't observe any improvement in performance. The best performance using both BERT and ULMFit was obtained without using any undersampling or oversampling. Therefore, the model that we used on the test data was trained on the full training dataset maintaining the given ratio of ADR:non-ADR tweets.

4.2 Task 2: Extraction of Adverse Effect Mentions

Model	Relaxed F1	Relaxed Precision	Relaxed Recall	Strict F1	Strict Precision	Strict Recall
BLSTM+CRF Tagger with Stacked Pretrained BERT and Twitter Glove Embeddings	0.638	0.532	0.796	0.315	0.262	0.395
BLSTM+CRF Tagger with Stacked Pretrained BERT and Flair Embeddings	**0.641**	0.537	0.793	**0.328**	0.274	0.409

Table 3: Results for Task-2: Extracting spans of text expressing adverse drug reactions in Tweets

Table 3, presents the performance scores for sub-task 2 on the test data. The different metrics as presented in the table were implemented by the organizers and the scores were provided by them.

4.3 Task 4: Generalized Identification of Personal Health Experience Mentions

The objective of the task is to classify whether a tweet contains a personal mention of ones health (for example, sharing ones own health status or opinion), as opposed to a more general discussion of the health issue, or an unrelated mention of the word. Each model was finally evaluated using four F1-scores - F1 for the held out influenza

Models	Accuracy	F1	Precision	Recall
Model 1 - BERT	0.8105	0.7453	0.9875	0.5985
Model 2 - BERT + Twitter Glove Embeddings	**0.8211**	**0.783**	0.8932	0.697
Model 3 - BERT + Flair Embeddings	0.8035	0.7544	0.8958	0.6515
Health Concerns Condition 1				
Model 1 - BERT	**0.9**	**0.8919**	1	0.8049
Model 2 - BERT + Twitter Glove Embeddings	0.8875	0.88	0.9706	0.8049
Model 3 - BERT + Flair Embeddings	0.8938	0.8859	0.9851	0.8049
Health Concerns Condition 2				
Model 1 - BERT	0.6377	0.359	0.875	0.2258
Model 2 - BERT + Twitter Glove Embeddings	**0.6667**	**0.5818**	0.6667	0.5161
Model 3 - BERT + Flair Embeddings	0.6087	0.4706	0.6	0.3871
Health Concerns Condition 3				
Model 1 - BERT	0.7679	0.48	1	0.3158
Model 2 - BERT + Twitter Glove Embeddings	**0.8214**	**0.6667**	0.9091	0.5263
Model 3 - BERT + Flair Embeddings	0.7857	0.5714	0.8889	0.4211

Table 4: Results for Task-4: Generalized identification of personal health experience mentions

data, the second and third undisclosed context, and the F1-score overall. The results that our models obtained on the test data is presented in Table 4. As already mentioned that the BLSTM models trained using pretrained embeddings could not be completed. Inspite of the fully trained model, we do see a decent performance using BLSTM along with a combination of pretrained embeddings on the provided dataset.

5 Future Work and Conclusion

In this work, we presented our initial attempt to use BERT and ULMFit for text classification tasks related to the domain of pharmacovigilance. We obtained decent results for three different tasks organized as a shared task in Social Media Mining for Health Workshop - 2019. We noticed that the BERT and ULMFit were agnostic to undersampling and oversampling unlike previously observed performances on traditional text classifiers as reported on a similar task (Sarker et al., 2018), that was a part of the same workshop held in 2017. We consider our reported work in this paper as a preliminary attempt and would like to extend them in the future. As part of our future work we would like to train better models using BERT for all the three sub-tasks that we participated in, and would also like to interpret the predictions of the models. We think domain specific training of different embeddings could help and would like to try them in the future.

References

Eiji Aramaki, Sachiko Maskawa, and Mizuki Morita. 2011. Twitter catches the flu: detecting influenza epidemics using twitter. In *Proceedings of the conference on empirical methods in natural language processing*, pages 1568–1576. Association for Computational Linguistics.

Munmun De Choudhury, Michael Gamon, Scott Counts, and Eric Horvitz. 2013. Predicting depression via social media. In *Seventh international AAAI conference on weblogs and social media*.

Jacob Devlin, Ming-Wei Chang, Kenton Lee, and Kristina Toutanova. 2018. Bert: Pre-training of deep bidirectional transformers for language understanding. *arXiv preprint arXiv:1810.04805*.

Kathleen M Giacomini, Ronald M Krauss, Dan M Roden, Michel Eichelbaum, Michael R Hayden, and Yusuke Nakamura. 2007. When good drugs go bad. *Nature*, 446(7139):975.

Jeremy Howard and Sebastian Ruder. 2018. Universal language model fine-tuning for text classification. *arXiv preprint arXiv:1801.06146*.

Alex Lamb, Michael J Paul, and Mark Dredze. 2013. Separating fact from fear: Tracking flu infections on twitter. In *Proceedings of the 2013 Conference of the North American Chapter of the Association for Computational Linguistics: Human Language Technologies*, pages 789–795.

Robert Leaman, Laura Wojtulewicz, Ryan Sullivan, Annie Skariah, Jian Yang, and Graciela Gonzalez. 2010. Towards internet-age pharmacovigilance: extracting adverse drug reactions from user posts to health-related social networks. In *Proceedings of the 2010 workshop on biomedical natural language processing*, pages 117–125. Association for Computational Linguistics.

Debanjan Mahata, Jasper Friedrichs, Rajiv Ratn Shah, and Jing Jiang. 2018a. Detecting personal intake of medicine from twitter. *IEEE Intelligent Systems*, 33(4):87–95.

Debanjan Mahata, Jasper Friedrichs, Rajiv Ratn Shah, and Jing Jiang. 2018b. Did you take the pill?-detecting personal intake of medicine from twitter. *arXiv preprint arXiv:1808.02082*.

Debanjan Mahata, Jasper Friedrichs, Rajiv Ratn Shah, et al. 2018c. # phramacovigilance-exploring deep learning techniques for identifying mentions of medication intake from twitter. *arXiv preprint arXiv:1805.06375*.

Puneet Mathur, Meghna Ayyar, Sahil Chopra, Simra Shahid, Laiba Mehnaz, and Rajiv Shah. 2018. Identification of emergency blood donation request on twitter. In *Proceedings of the 2018 EMNLP Workshop SMM4H: The 3rd Social Media Mining for Health Applications Workshop & Shared Task*, pages 27–31.

Atul Nakhasi, Ralph Passarella, Sarah G Bell, Michael J Paul, Mark Dredze, and Peter Pronovost. 2012. Malpractice and malcontent: Analyzing medical complaints in twitter. In *2012 AAAI Fall Symposium Series*.

Yishai Ofran, Ora Paltiel, Dan Pelleg, Jacob M Rowe, and Elad Yom-Tov. 2012. Patterns of information-seeking for cancer on the internet: an analysis of real world data. *PloS one*, 7(9):e45921.

Michael J Paul and Mark Dredze. 2011. You are what you tweet: Analyzing twitter for public health. In *Fifth International AAAI Conference on Weblogs and Social Media*.

Abeed Sarker, Maksim Belousov, Jasper Friedrichs, Kai Hakala, Svetlana Kiritchenko, Farrokh Mehryary, Sifei Han, Tung Tran, Anthony Rios, Ramakanth Kavuluru, et al. 2018. Data and systems for medication-related text classification and concept normalization from twitter: insights from the social media mining for health (smm4h)-2017 shared task. *Journal of the American Medical Informatics Association*, 25(10):1274–1283.

Abeed Sarker and Graciela Gonzalez. 2015. Portable automatic text classification for adverse drug reaction detection via multi-corpus training. *Journal of biomedical informatics*, 53:196–207.

Smaranda Muresan Tuhin Chakrabarty. 2019. Columbianlp at semeval-2019 task 8: The answer is language model fine-tuning. In *Proceedings of the 12th International Workshop on Semantic Evaluation*, pages 1140–1144.

Davy Weissenbacher, Abeed Sarker, Arjun Magge, Ashlynn Daughton, Karen O'Connor, Michael Paul, and Graciela Gonzalez-Hernandez. 2019. Overview of the fourth Social Media Mining for Health (SMM4H) shared task at ACL 2019. In *Proceedings of the 2019 ACL Workshop SMM4H: The 4th Social Media Mining for Health Applications Workshop Shared Task*.

Amir Hossein Yazdavar, Hussein S Al-Olimat, Monireh Ebrahimi, Goonmeet Bajaj, Tanvi Banerjee, Krishnaprasad Thirunarayan, Jyotishman Pathak, and Amit Sheth. 2017. Semi-supervised approach to monitoring clinical depressive symptoms in social media. In *Proceedings of the 2017 IEEE/ACM International Conference on Advances in Social Networks Analysis and Mining 2017*, pages 1191–1198. ACM.

Detection of Adverse Drug Reaction in Tweets Using a Combination of Heterogeneous Word Embeddings

Segun Taofeek Aroyehun
CIC, Instituto Politécnico Nacional
Mexico City, Mexico
aroyehun.segun@gmail.com

Alexander Gelbukh
CIC, Instituto Politécnico Nacional
Mexico City, Mexico
www.gelbukh.com

Abstract

This paper details our approach to the task of detecting reportage of adverse drug reaction in tweets as part of the 2019 social media mining for healthcare applications shared task. We employed a combination of three types of word representations as input to a LSTM model. With this approach, we achieved an F1 score of 0.5209.

1 Introduction

The social media mining for health care applications shared task aims to provide a benchmark for validating and comparing methods for healthcare applications using social media data (Weissenbacher et al., 2019). The focus of task 1 is on identifying adverse drug reaction as a medication related outcome. Participants on this task are expected to differentiate tweets as reporting adverse drug reaction or not and the performance metric is F1. This task demands that adverse drug reaction be distinguished from a similar and mostly confounding expression of the indication of a drug. The former is usually associated with the usage of the drug while the latter is a specification of the reason to use a drug. In addition, the task of detecting mention of adverse drug reaction is an extremely imbalanced binary classification task. About 1% of the training set are positive examples and approximately 99% are negative examples. Our approach is based on the combination of three different types of word embedding representations viz: character (Lample et al., 2016), non-contextual(Glove pre-trained on Twitter data) (Pennington et al., 2014), and contextual(BERT) (Devlin et al., 2018). The following section gives details of our model and training set-up. Section 3 shows the results of our experiments while we conclude and speculate on future directions in Section 4.

2 Model and Experimental Set-Up

We hypothesize that the different types of embeddings capture different relationships and their combination could help in the identification of adverse drug reaction in tweets. In our experiments, the word representation differs in two dimensions: whether they are pre-trained (Glove and Bert) or not (character embedding) and if they are contextual (Bert) or otherwise (Glove and Character embeddings). We briefly describe each representation:

- Character embedding - is a 50 dimensional representation of the characters in a word (how are they combined to form an embedding for the word). This representation is trained together with the model. It is based on a bidirectional LSTM. The advantage of character-based representation for social media text is that it eliminates the out-of-vocabulary problem which results from noise in the form of misspellings and abbreviations in word-based representation such as Glove. Also, this representation is specific to the task and domain of the training set.

- Glove (twitter) - is a 100 dimensional representation pre-trained on a huge twitter corpus. We expect this to contribute by reflecting the language of twitter users. However, the embedding is not a contextual one.

- BERT (en, base-uncased) - is a general domain contextual word representation where the representation of a word is based on other words in its context (sentence). The BERT base model which is not cased gives a word embedding of dimension 768. It has enabled state-of-the-art results on several NLP tasks. However, to the best of our knowledge, its application to social media text is limited.

Proceedings of the 4th Social Media Mining for Health Applications (#SMM4H) Workshop & Shared Task, pages 133–135
Florence, Italy, August 2, 2019. ©2019 Association for Computational Linguistics

	No. of Examples
train	24202
dev	6051
test	4575

Table 1: Details of the Data

	P	R	F1
sub1	0.6145	0.4457	0.5167
sub2	**0.6203**	**0.4489**	**0.5209**

Table 2: Performance on the Test Set (Scores as provided by the organizers)

Model	F1
emb comb w/ fine tuning	0.9015
emb comb w/o fine tuning	0.9060
emb comb w/ fine tuning w/o character	0.8777
emb comb w/ fine tuning w/o Glove	0.9020
emb comb w/ fine tuning w/o BERT	0.9040

Table 3: Performance of Model Variants on the Validation Split

In order to leverage some of the benefits of the representations above, we concatenated these representations for a given word in a tweet. This combination is of dimension 918. A linear layer then project this representation into a dimension of 256. This projection is meant to serve as a distillation step and/or as a fine-tuning step. The resulting representation is fed into an LSTM layer with hidden size of 512 to sequentially model a tweet. Finally, a dense layer is used as the classifier.

The model was trained for 100 epochs with learning rate annealing factor of 0.5 using SGD as the optimizer and a batch size of 8. We used a train-dev split of 80:20. Table 1 shows the number of training, validation, and evaluation examples used in our experiment. Weissenbacher et al. (2019) provide details on the collection and annotation of the dataset. Based on the validation split, a model with the best F1 score is saved during training as the best model. With the best model, we made predictions on the unseen evaluation examples as our first submission (sub1 in Table 2). Our second submission (sub2 in Table 2) was based on the model at the 100th epoch or the last epoch as training is terminated if learning rate becomes too small. Our experiments were performed using the Flair framework (Akbik et al., 2018).

3   Results

Table 3 shows the results obtained on the test set. We achieved our best submission with the final model with an F1 of 0.5209. This result ranks above the average score of all participants in the task with average F1, precision, and recall of 0.5019, 0.5351, 0.5054 respectively (Weissenbacher et al., 2019). Table 3 shows the results obtained from our ablation experiments with respect to the contributions of the different embedding representations and the distillation/fine-tuning step. The F1 scores reported are based on the model that achieved the best F1 score on the validation set during training. We observed a minimal drop in performance (0.0045) when we removed the fine-tuning layer. This suggests that the fine-tuning layer either hurts performance or the dimension of the resulting fine-tuned representation is an important parameter to tune with our approach. We assessed the contribution of the three embedding representations to performance by removing one at a time from the model while keeping our fine-tuning strategy. When the character embedding word representation is absent, a performance drop of 0.0238 is observed. When the BERT representation is removed, the performance improved by 0.0025. Without the Glove embedding, the performance increased by 0.0005. This result is consistent with our perceived advantages and disadvantages of the three embedding representations. With the character embedding contributing the most to the model performance. Remarkably, the removal of BERT and Glove leads to improved performance. This can be attributed to the out-of-vocabulary problem with Glove and domain mismatch in the case of BERT.

4   Conclusion

This paper outlines our participation in the 2019 social media mining for healthcare application challenge on identifying the reportage of adverse drug reaction in tweets. Our approach is based on the combination of three different types of embedding representations and a fine-tuning strategy. With this approach, we made two submissions using a model that achieved the best F1 score on the validation data and with a model trained till the last epoch possible. The latter gave a better performance. Through ablation experiments, we observed that our fine-tuning strategy results in a small drop in performance contrary to our expectation. In addition, the different word representations contribute to different degrees. The character embedding representation makes the most significant contribution, without it the model performance drops while there is a marginal performance improvement when both Glove and BERT representation are removed from the model.

As a follow-up work, we would like to investigate other fine-tuning or distillation approaches as well as parameter tuning of the size of the fine-tuning layer. It is also interesting to examine the impact of normalizing tweets and identifying usage expressions as an auxiliary task.

References

Alan Akbik, Duncan Blythe, and Roland Vollgraf. 2018. Contextual string embeddings for sequence labeling. In *COLING 2018, 27th International Conference on Computational Linguistics*, pages 1638–1649.

Jacob Devlin, Ming-Wei Chang, Kenton Lee, and Kristina Toutanova. 2018. Bert: Pre-training of deep bidirectional transformers for language understanding. *arXiv preprint arXiv:1810.04805*.

Guillaume Lample, Miguel Ballesteros, Sandeep Subramanian, Kazuya Kawakami, and Chris Dyer. 2016. Neural architectures for named entity recognition. In *Proceedings of NAACL-HLT*, pages 260–270.

Jeffrey Pennington, Richard Socher, and Christopher Manning. 2014. Glove: Global vectors for word representation. In *Proceedings of the 2014 conference on empirical methods in natural language processing (EMNLP)*, pages 1532–1543.

Davy Weissenbacher, Abeed Sarker, Arjun Magge, Ashlynn Daughton, Karen O'Connor, Michael Paul, and Graciela Gonzalez-Hernandez. 2019. Overview of the Fourth Social Media Mining for Health (SMM4H) Shared Task at ACL 2019. In *Proceedings of the 2019 ACL Workshop SMM4H: The 4th Social Media Mining for Health Applications Workshop & Shared Task*.

Identification of Adverse Drug Reaction Mentions in Tweets – SMM4H Shared Task 2019

Samarth Rawal **Siddharth Rawal** **Saadat Anwar** **Chitta Baral**

Arizona State University Arizona State University Arizona State University Arizona State University

samrawal@asu.edu sidrawal@asu.edu sanwar@asu.edu chitta@asu.edu

Abstract

Analyzing social media posts can offer insights into a wide range of topics that are commonly discussed online, providing valuable information for studying various health-related phenomena reported online. The outcome of this work can offer insights into pharmacovigilance research to monitor the adverse effects of medications. This research specifically looks into mentions of adverse drug reactions (ADRs) in Twitter data through the Social Media Mining for Health Applications (SMM4H) Shared Task 2019. Adverse drug reactions are undesired harmful effects which can arise from medication or other methods of treatment. The goal of this research is to build accurate models using natural language processing techniques to detect reports of adverse drug reactions in Twitter data and extract these words or phrases.

1 Introduction

On average, one in a thousand messages from public Twitter data is health-related (Sadilek et al., 2012). These health-related Twitter posts can be used to monitor and analyze various health-related phenomena such as drug use and side effects resulting from medication. The purpose of this work was to develop a model to accurately analyze mentions of adverse drug reactions (ADRs) in Twitter posts. To achieve this task, natural language processing techniques were used to predict whether each Tweet from a given set of Tweets contains a mention of an ADR and extract any mentions of ADRs. The results of this project can be useful for research done in the field of pharmacovigilance, which is the monitoring of drug effects with the intention of finding and preventing adverse effects. This work was conducted as part of the Social Media Mining for Health (SMM4H) challenge hosted by the Health Language Processing (HLP) Lab at the University of Pennsylvania. The predictions of the models developed for this project were evaluated against test data and given F-scores as well as scores of accuracy, precision, and recall based on the degree to which they were able to accomplish the goals of each task.

2 Methods

2.1 Subtask 1

For Subtask 1, a lexicon-based approach was followed. To identify important keywords – keywords whose presence or absence in a Tweet can serve as valuable, reliable indicators of whether the Tweet contains a reference to an Adverse Drug Reaction or not – a methodology adapted from the Internal + External Lexicon Selection technique (Rawal et al., 2019), a technique that has yielded successful results in previous similar classification tasks, was used. First, uni- and bi-grams were extracted from the training dataset. The presence or absence of each of these n-grams were then used as binary features in a logistic regression model. To estimate the performance of the model using metrics that were to be used for evaluation, such as precision, recall, and F1-score, the model was trained via 10-fold cross-validation of the training set. Finally, the coefficients associated with each keyword were examined. There were 166,466 total features obtained through the aforementioned technique. Through this process, the top 700 absolute-valued coefficients were hypothesized to be the most significant keywords and stored. This number of top keywords to keep was a hyperparameter that was experimentally determined through model performance over 10-fold cross-validation of the training set. This list of significant keywords was then manually pared down to exclude any intuitively irrelevant terms (such as stop words); the presence or absence of these remaining keywords were used as binary features for our final logistic regression model. Other mod-

Proceedings of the 4th Social Media Mining for Health Applications (#SMM4H) Workshop & Shared Task, pages 136–137
Florence, Italy, August 2, 2019. ©2019 Association for Computational Linguistics

els were also tested during training, such as a BioBERT (Lee et al., 2019) model that was fine-tuned using the provided training data. Although the BioBERT model showed promising results, it was not implemented into the final submission due to time constraints.

2.2 Subtask 2

For Subtask 2, a deep learning approach was taken. Specifically, a Bidirectional Long Short-Term Memory (BiLSTM) coupled with a Conditional Random Field (CRF) layer neural network architecture was used to perform Named Entity Recognition to identify the Adverse Drug Reaction mentions. This architecture has been empirically shown to perform well at Named Entity Recognition (NER) tasks (Lample et al., 2016). To represent input words, the Embedding layer weights of the model was pre-initialized with values obtained from a word2vec model that was trained on the MIMIC-III dataset (Johnson et al., 2016).

Figure 1: BiLSTM-CRF neural network architecture

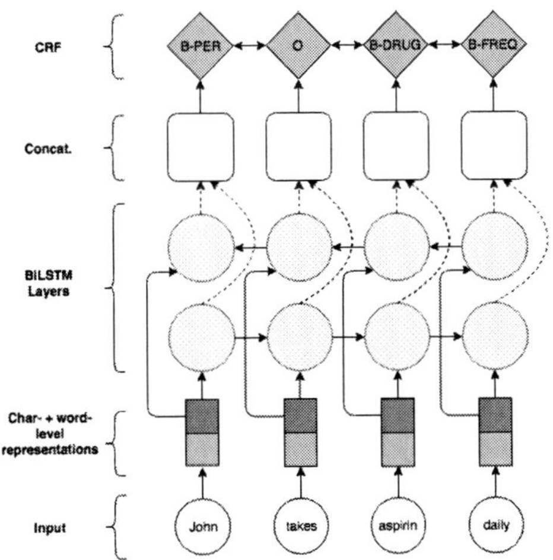

3 Results

On Task 1, our system performed with an F1 score of 0.4317, Precision of 0.3223, and Recall of 0.6534.

On Task 2, on the relaxed metric, our system performed with an F1 score of 0.535, Precision of 0.415, and Recall of 0.753; on the strict metric,

our system performed with an F1 score of 0.269, Precision of 0.206, and Recall of 0.390.

4 Conclusion

Overall, our systems for Tasks 1 and 2 consisted of a combination of (1) lexicon selection and domain-specific feature engineering; (2) classical machine learning techniques such as logistic regression; and (3) neural architectures, including BioBERT and BiLSTM-CRF models. We found simpler models consisting of lexicon selection and classical machine learning models (such as the logistic regression model discussed previously) performed better with limited datasets and offered explainability into feature importance. In the Named Entity Recognition task, we utilized a deep learning approach, given the demonstrated effectiveness of such an architecture in this domain (Lample et al., 2016). We expect to improve the performance of our systems through further refinement of our feature engineering and tuning of our model parameters.

Acknowledgments

The research team would like to thank Dr. Chitta Baral of Arizona State University for providing guidance and mentorship for this project.

References

Alistair EW Johnson, Tom J Pollard, Lu Shen, H Lehman Li-wei, Mengling Feng, Mohammad Ghassemi, Benjamin Moody, Peter Szolovits, Leo Anthony Celi, and Roger G Mark. 2016. Mimic-iii, a freely accessible critical care database. *Scientific data*, 3:160035.

Guillaume Lample, Miguel Ballesteros, Sandeep Subramanian, Kazuya Kawakami, and Chris Dyer. 2016. Neural architectures for named entity recognition. *arXiv preprint arXiv:1603.01360*.

Jinhyuk Lee, Wonjin Yoon, Sungdong Kim, Donghyeon Kim, Sunkyu Kim, Chan Ho So, and Jaewoo Kang. 2019. Biobert: a pre-trained biomedical language representation model for biomedical text mining. *CoRR*, abs/1901.08746.

Samarth Rawal, Ashok Prakash, Soumya Adhya, Sidharth Kulkarni, Saadat Anwar, Chitta Baral, and Murthy V. Devarakonda. 2019. Developing and using special-purpose lexicons for cohort selection from clinical notes. *CoRR*, abs/1902.09674.

Adam Sadilek, Henry Kautz, and Vincent Silenzio. 2012. Modeling spread of disease from social interactions.

Author Index

Association for Computational Linguistics
209 N. Eighth Street
Stroudsburg, Pennsylvania 18360

ISBN 978-1-5108-9113-5